NOT ALL DEAD WHITE MEN

NOT ALL
DEAD
WHITE
MEN

Classics and Misogyny in the Digital Age

DONNA
ZUCKERBERG

Harvard University Press

CAMBRIDGE, MASSACHUSETTS   LONDON, ENGLAND   2018

Library of Congress Cataloging-in-Publication Data

Names: Zuckerberg, Donna, 1987– author.
Title: Not all dead white men : classics and misogyny in the digital age /
    Donna Zuckerberg.
Description: Cambridge, Massachusetts : Harvard University Press, 2018. |
    Includes bibliographical references and index.
Identifiers: LCCN 2018013357 | ISBN 9780674975552 (alk. paper)
Subjects: LCSH: Right-wing extremists—United States—History—
    21st century. | Men's movement—United States—History—21st century. |
    Misogyny—United States—History—21st century. |
    United States—Civilization—Classical influences.
Classification: LCC HN49.R33 Z83 2018 | DDC 320.53—dc23
LC record available at https://lccn.loc.gov/2018013357

CONTENTS

NOT ALL DEAD WHITE MEN

INTRODUCTION

At the end of 2016, posters for the white nationalist group Identity Evropa began to appear on college campuses in the United States. The posters featured black-and-white photographs of statues, most of which were either ancient, such as the Apollo Belvedere, or obviously classicizing, such as Nicolas Coustou's 1696 statue of Julius Caesar.[1] Overlaid on these images were generic, seemingly inoffensive slogans such as "Protect Our Heritage" and "Our Future Belongs to Us." The posters caused a wave of outrage and were quickly removed, although they remained available for sale on the Identity Evropa website under the heading "Epic Posters" for nearly a year.

This use of classical imagery to promote a white nationalist agenda is far from an isolated occurrence. In fact, the Identity Evropa posters are unusual not for what they depict but, rather, for having an actual physical presence. In the less tangible world of the internet, far-right communities ideologically aligned with Identity Evropa have increasingly been using artifacts, texts, and historic figures evocative of ancient Greece and Rome to lend cultural weight to their reactionary vision of ideal white masculinity.

These online communities go by many names—the Alt-Right, the manosphere, Men Going Their Own Way, pickup artists—and exist under the larger umbrella of what is known as the Red Pill, a group of men connected by common resentments against women, immigrants, people of color, and the liberal elite. The name, adopted from the film *The Matrix*, encapsulates the idea that society is unfair to men—heterosexual white men in particular—and is designed to favor women. The Red Pill finds its primary online home on the subreddit r/theredpill, a forum on the social media platform Reddit dedicated to discussion of Red Pill ideas. Its influence and reach, however, extend far beyond that home: men in Red Pill communities—on Reddit and elsewhere online—share articles, memes, and news stories to incite one another's anger. That anger then occasionally finds outlets in what are sometimes called troll storms: a hurricane of digital abuse aimed at those with the misfortune to attract attention.

The Red Pill community has an odd and uncomfortable relationship with social media: its members exhibit widespread disdain for every major social media platform, but they also use those platforms as major modes of communication and object vociferously when members of the community are banned from social media sites. James "Roissy" Weidmann, writer of the popular blog *Chateau Heartiste*, calls Twitter "Twatter," and *Return of Kings*, a popular blog within the manosphere community, frequently publishes articles arguing that Twitter's censorship of conservative personalities such as Milo Yiannopoulos will lead to its eventual bankruptcy. Many members of the community have a presence on both Twitter and Gab, a less restrictive Twitter clone, and some factions of the community have relocated from the barely policed news aggregator Reddit to its even less restrictive counterpart Voat. Mark Zuckerberg, the founder of Facebook (and my older brother), is frequently mocked as "Mark Cuckerberg" or "Zuck the Cuck," epithets based on the term *cuck*, a particularly significant form of insult within the Red Pill derived from the term *cuckold*.

I understand what it feels like to have an ambivalent relationship with social media. I moved to Silicon Valley in 2012, when my husband accepted a job at a social media marketing company that was later acquired by Google. All three of my siblings have worked in social media, and so have many people in my social circle. Because

Introduction

I know so many people working in the technology industry, I hear a great deal about the power of technology to connect the world and build communities. But when people with similar interests are connected, some of the strengthened communities will inevitably be those bound by shared hatreds and prejudices. The communities studied in this book are a perfect example. Social media has led to an unprecedented democratization of information, but it has also created the opportunity for men with antifeminist ideas to broadcast their views to more people than ever before—and to spread conspiracy theories, lies, and misinformation. Social media has elevated misogyny to entirely new levels of violence and virulence.

Anyone today who does not intend to become a digital hermit is guaranteed to encounter these men online. Those inevitable encounters will be less traumatic and shocking to those who are prepared and able to recognize the strategies they use to attack their targets—including how they use Greek and Roman antiquity to bolster their credibility.

The Red Pill community is by no means unique in its attraction to ancient Greece and Rome. Political and social movements have long appropriated the history, literature, and myth of the ancient world to their advantage. Borrowing the symbols of these cultures, as the Nazi Party did in the 1940s, can be a powerful declaration that you are the inheritor of Western culture and civilization.[2] The men of the Red Pill have adapted this strategy for the digital age. They have turned the ancient world into a meme: an image of an ancient statue or monument becomes an endlessly replicable and malleable shorthand for projecting their ideology and sending it into the world.

Classics is not the only field of inquiry these men use to justify their views. They are particularly interested in the histories of Great Britain, Germany, and Russia, especially the medieval period, and they also compose and cite articles about evolutionary psychology, philosophy, biology, and economics. The Greek and Roman Classics,

3

nonetheless, hold particular cultural significance for them. By turning frequently to authors such as Marcus Aurelius and Ovid, they attempt to perpetuate the idea that white men are the guardians of intellectual authority, especially when such authority is perceived to be under threat from women and people of color. They claim that the ancient world and, by extension, the study of the ancient world are under attack by the "politically correct establishment" and "social justice warriors" in US classrooms. As colleges move to replace some of the dead white men of the literary canon with writers who are not dead, not white, and not men, the living white men of the Red Pill have appeared as the self-appointed guardians and defenders of the cultural legacy of Western civilization.

Red Pill engagement with the Classics would be concerning even if it were simply a matter of a few internet trolls writing for an audience of a few hundred thousand more internet trolls. These men, no matter how small their numbers, have a disproportionately loud presence in the online discourse about sex and gender, and it would be necessary to explore how they use antiquity to construct their authority. Unfortunately, however, the far-right abuse of Classics extends beyond just a few online publications and subreddits.

The election of President Donald Trump in 2016 empowered these online communities to be even more outspoken about their ideology. As one manosphere thought-leader wrote, "His presence [in office] automatically legitimizes masculine behaviors that were previously labeled sexist and misogynist"—but, of even greater concern, it also put a few men who share those ideas into positions of power near the president.[3] Steve Bannon, the former White House chief strategist and, earlier, the executive chair of the far-right website *Breitbart News* (which he once famously called "the platform for the Alt-Right"), is a lover of the Classics; one screenwriter who worked with Bannon—on a hip-hop version of Shakespeare's *Coriolanus*—recalls that "he was always quoting [Marcus] Aurelius."[4] And Michael Anton, a national security official in the Trump administration, wrote essays in *The Claremont Review* and other websites during the election under the pseudonym Publius Decius Mus, after a fourth-century BCE Roman consul.[5] Those essays would end up providing an intellectual foundation for Trumpism, which Anton defined in his essay "The Flight 93 Election" as "secure borders, economic nationalism, and America-first foreign policy."[6] Those who frequent Red Pill message boards have embraced these two men as heroes.

It would be an exaggeration to say that the men of the Red Pill community are writing national policy. However, on some level, they seem to *believe* they are influencing policy, and that belief has empowered them. Their numbers are also swelling: as of this writing, the subreddit r / theredpill has over 230,000 subscribers, up from 138,000 at the beginning of 2016. The members of this growing community are more confident than ever that their gender- and race-based politics are validated both by science and by the Western tradition, and they believe that highly placed members of the Trump administration agree with them.

———

This book is about how the men of the Red Pill use the literature and history of ancient Greece and Rome to promote patriarchal and white supremacist ideology. My goal is to lay bare the mechanics of this appropriation: to show how classical antiquity informs the Red Pill worldview and how these men weaponize Greece and Rome in service of their agenda. Anybody who has an interest in the Classics or social justice should not ignore this trend, which has the potential to reshape what ancient Greece and Rome mean in the twenty-first century while simultaneously promoting dangerous and discriminatory views about gender and race.

I have decided to focus primarily on the gender politics rather than the racial politics of Red Pill communities for two reasons.

First, the gender politics are generally more coherent throughout the Red Pill, with a shared interest in policing the sexuality and reproduction of young women (particularly young white women), whereas outright white supremacy is a more hotly contested issue within the community.[7] Second, the use of the ancient world to understand gender and sex is bidirectional: the men of the manosphere see their own misogyny reflected back at them, theorized, and celebrated in ancient literature. White supremacy is less easy to retroject onto the ancient world, which had no meaningful concept of biological race, as many scholars have shown.[8] But although whiteness is not a meaningful concept to apply to antiquity, that conceptual lacuna has not stopped the Alt-Right from using ancient Greece and Rome to fabricate a cohesive transhistorical "white" identity and a continuity of "European" or "Western" civilization for themselves. It has, however, kept them from using ancient literature to help them theorize whiteness, as the manosphere has done with masculinity, and thus their discourse about ancient race is necessarily more superficial.[9]

This book is written for people who have an interest in Classics but have not studied it extensively. It does not, therefore, contain an extensive social history about the lives of women in ancient Greece and Rome. I provide context for the ancient texts and historical figures under discussion throughout, as well as some basic background in Chapter 4 on sexual violence in the ancient world. For readers interested in learning more about women in antiquity, I offer suggestions for further reading in the endnotes. I also do not focus on the history of aggrieved masculinity in America, as social historians such as Michael Kimmel have done. The problems faced by men in America—the problems the Red Pill community points to as proof that we live in a gynocentric society—have deep historical roots; however, the Red Pill represents a new, dangerous phase of American masculinity in the internet age.[10]

Although my focus in this book is on the use of classical antiquity in the Red Pill community, this relatively narrow topic can provide a path into a deeper understanding of the Red Pill community as a whole. In researching this book, I spent years reading articles, posts, and comment sections on Red Pill websites large and small, on a range of topics from professional success to personal fitness to relationship advice. I have drawn my examples primarily from the most read sites and most influential thought leaders within the Red Pill—the writers with many followers and the articles with many comments.

The first chapter of this book describes the various factions within the Red Pill in greater detail and explains why the ancient world holds such appeal for them. Despite the movement's apparent disorganization and undertheorized positions, I argue that we would be wise to pay attention to them rather than dismissing them as an impotent fringe movement. The need to take the Red Pill seriously is especially urgent for feminists who use the internet and social media for personal and professional communications. As soon as a woman self-identifies online as a feminist, she is likely to find herself in a hailstorm of abusive tweets and emails from the men who frequent Red Pill websites. Understanding their ideology and tactics for online intimidation can help lessen the impact of that abuse.

In Chapter 2, I explore the fascination with ancient Stoicism displayed on Red Pill websites, which frequently discuss Stoic ideas and texts. In particular, writers use Stoicism to justify their belief that women and people of color are not just angrier and more emotional than men, but morally inferior as well. The growing community of Stoicism enthusiasts outside of the Red Pill, I argue, should not simply dismiss this use of the philosophy; instead, it should seek to understand how Stoicism's tenets can lend themselves to the perpetuation of systemic injustice.

Not All Dead White Men

Chapter 3 examines one particular faction of the manosphere: the community of pickup artists who claim the Roman poet Ovid as the first person to write a seduction manual. The *Ars Amatoria*, written over two thousand years ago, is a fascinating and contradictory work that has puzzled Latin scholars with its playful tone and apparent justifications for sexual assault. Reading Ovid alongside the advice of pickup artists offers us insight into seduction methodology in both ancient and modern times: both rely on the ideas that women's boundaries are permeable and consent is a flexible concept.

While the third chapter focuses on the sexual politics of the Red Pill in the contemporary world, the fourth and final chapter addresses how ancient literature informs their *aspirational* sexual politics: how, in their ideal world, men and women would interact. This ideal patriarchy draws heavily on ancient models of marriage and family to promote a world in which women have no decision-making power outside of the home. This chapter also addresses the Red Pill fixation with false rape allegations, one of the most popular topics on many Red Pill fora and their ultimate proof that we live in a society where women have more privilege than men. I use the myth of Phaedra and Hippolytus, an ancient example of a false allegation with disastrous consequences, to show that, because there was anxiety about false allegations in the deeply patriarchal ancient world as well, the Red Pill use of the trope is actually a tool for misdirection. These men not only wish to prevent false allegations from occurring; they also wish to resurrect a world where female consent to sexual activity is as negligible a concern as it was in the ancient world.

As Angela Nagle argues in her 2017 book *Kill All Normies: Online Culture Wars from 4chan and Tumblr to Trump and the Alt-Right*, far-right internet subcultures are powered by a politics of transgression.[11] It is perhaps ironic that the ancient Greek and

Roman Classics, with all of their considerable, well-recognized cultural capital, have been embraced so vocally by a fundamentally countercultural movement. The men of the Red Pill use their vision of an idealized version of Western civilization and its past to critique our own society and inspire change, and they often rhetorically position this strategy as the natural, obvious way to understand what classical antiquity means in the present day. But by analyzing and deconstructing this Red Pill enthusiasm for ancient Greece and Rome, I hope to articulate a different vision for a feminist, radical place that classical antiquity can occupy in contemporary political discourse.

As classical scholars have become aware of the Far Right's appropriation of antiquity, some have responded by suggesting that we should focus on pointing out how inexpert these appropriations tend to be and how little actual knowledge of ancient Greece and Rome they reveal.[12] But while Red Pill references to the Classics are often inaccurate, confounding, or lacking in nuance, they can be dangerous nevertheless. Even the most elementary errors still leverage the ancient world to promote reactionary ideas about gender and race. I do not, therefore, devote too much energy to correcting flawed Red Pill classical interpretations. I do occasionally point out the most blatant errors in order to avoid perpetuating misinformation about the classical world, but identifying such errors is far from an end in itself. By focusing on the Red Pill community's flawed understanding of ancient Greece and Rome, scholars may miss opportunities to engage with the deeper ideological purpose of classical appropriation.

Marcus Aurelius, one of the Red Pill community's favorite ancient writers, once wrote, "It's ridiculous to try to escape other people's flaws and not your own—to try the impossible rather than the possible" (*Meditations* 7.71). We cannot stop these men from using and abusing the history and literature of the ancient world in service of a patriarchal, white nationalist agenda. But by revealing how this self-mythologizing works, we can develop strategies for counteracting its pernicious influence.

1

ARMS AND THE MANOSPHERE

You may not have heard of the manosphere or the Red Pill before reading this book. If, however, you spend any time on the internet, you probably know the men who constitute these far-right, antifeminist online groups. These men are the ones coordinating attacks to send death and rape threats to outspoken feminists. They are a significant part of why the comments section in many online mainstream media articles is nearly unreadable. They believe it is their right and duty to invade feminist spaces. They are convinced that sexism—attitudes and behavior that foster discrimination against women and perpetuate gender-based stereotypes—is really a form of enlightenment and that they are the only logical people on the internet. Since some of these men are skilled at deploying emotional abuse tactics, they succeed surprisingly often at convincing people that their worldview has a rational basis.

These online communities connect a large group of cisgender men—that is, men whose identity aligns with the sex they were assigned at birth—united by the belief that masculine cisgender men are discriminated against by our feminized ("gynocentric") society and must support each other. A few self-reported surveys within the community suggest that more than three quarters of these men are white, heterosexual, politically conservative, have no strong religious affiliation, and are between the ages of eighteen and thirty-five.[1] Although there are similar men's movements in other countries, as well as similar far-right nationalist groups, these men are resolutely focused on the concerns of men in the United States.[2]

Instead of seeing themselves as part of the nation's most affluent and powerful demographic, the predominantly white heterosexual men of the Red Pill believe they need solidarity with each other because the idea of white male supremacy is an illusion maintained to ensure they remain oppressed. Although they concede that many of the most powerful people in the world are men—and are happy to use that as evidence that men are intellectually superior and more naturally suited to dominance and leadership than women are—they believe that the "myth of male privilege" is a manifestation of "the apex fallacy": the tendency to judge the status of an entire group based on a few outstanding members. Just as liberals would argue that the election of a black President does not mean racism is no longer a concern, these men argue that the fact that every commander-in-chief of the United States has been male does not signify that men are not in a relatively disadvantaged position in our society.

Some of their evidence that men in the United States are in unfavorable circumstances is compelling. Men are more likely than women to be victims of violence. Men represent more than 90 percent of the prison inmate population. In this country, and in almost every developed nation across the world, men commit suicide at a rate almost three times that of women. The vast majority of people killed in the workplace are men. Female students outnumber male students in primary, secondary, and college classes. And, unlike men, women are almost never falsely accused of rape or forced to pay child support for children over whom they have no parental rights.

If the men on Red Pill message boards truly focused on finding solutions to these problems or understanding their complex underlying causes, I would not have written this book. Unfortunately, instead of looking for answers, they prefer to "fight the cultural narrative." According to the Red Pill, the rise of feminism and progressivism is both the cause of the problems plaguing men and, through the "cultural narrative" they generate, the reason those problems are not taken seriously. They "fight" this narrative by looking for scapegoats: corporate America, liberals, immigrants, and—most of all—women, whom they harass and abuse both online and in person. The Red Pill community appears to embody toxic masculinity, which psychiatrist Terry Kupers, an expert on mental health in prisons, defines as "the constellation of socially regressive male traits that serve to foster domination, the devaluation of women, homophobia, and wanton violence."[3]

Indeed, the concept of an overarching narrative is central to how they construct their identity. Many men in the Red Pill describe a conversion process of discovering how unfair the world truly is to men, often called "swallowing the red pill"—a reference to the famous scene in the film *The Matrix* (1999) in which Morpheus (Laurence Fishburne) offers Neo (Keanu Reeves) a choice to return to blissful ignorance or learn the truth about their reality. Morpheus tells Neo, "You take the blue pill—the story ends, you wake up in your bed and believe whatever you want to believe. You take the red pill—you stay in Wonderland, and I show you how deep the rabbit hole goes. Remember: all I'm offering is the truth."[4] One of the largest fora for the men studied in this book is a community called "The Red Pill" (often shortened to TRP) on Reddit. As of this writing, the subreddit r/theredpill has over 230,000 members and r/mensrights over 160,000, although it is safe to guess that many users subscribe to both subreddits. There are also dozens of related subreddits, such as r/pussypassdenied, a community of 160,000 subscribers dedicated to promoting what they perceive as true gender equality by dismantling "female privilege."

It has been suggested to me that I too might want to create my own terminology for the Red Pill factions in this book rather than use the many difficult-to-keep-track-of acronyms and terms they adopt as their identity.[5] There are obvious advantages to this tactic.

It would have allowed me to be clearer and more systematic, since the emic terms are both confusing and plentiful. It would also avoid the risk of appearing to validate or normalize these groups by accepting them on their own terms—an argument many use when maintaining that the group calling itself the Alt-Right should instead be referred to as neo-Nazis or white nationalists. I have chosen to use primarily the internally approved self-designations, although I provide a glossary in the back of the book to help the reader keep track of the many acronyms that populate the Red Pill, because in this case, emic terminology highlights how these men construct online identities through ideology and community.[6] Being part of the Alt-Right is as much about being part of a specific movement and moment as it is about a philosophical alignment with white supremacy.[7] Identifying these movements and moments by the terms they themselves use is a crucial part of understanding why the Red Pill community resonates with so many men.

In this chapter, I will map out the contours of the Red Pill community and its various subgroups. By identifying what each of these internal factions is trying to achieve and how they operate, we can better understand why they feel compelled to position themselves as the inheritors of the classical tradition and how the ancient world validates one of their most cherished, deeply held beliefs: that all women throughout history share distinct, immutable qualities that make them promiscuous, deceitful, and manipulative.

Ingredients in the Red Pill

While trolls have been part of the internet landscape for decades, the Red Pill community as it exists today seems to have coalesced around 2012. It began as a group of websites with a self-conscious focus on men's issues, including the subreddit founded in 2012 by Robert Fisher—later a Republican politician—called r/theredpill.[8] Internet sexism predates that period, but the Red Pill represents a new phase in online misogyny. Its members not only mock and belittle women; they also believe that in our society, men are oppressed by women. Early Red Pill blogs and fora, including the currently defunct site The Spearhead, started as places for like-minded men to discuss the problems men face in the United States today—false rape allegations, fathers' rights, the unfairness of the dating market. Within a few years, however, the community began to split into distinct, adversarial units.

Because the Red Pill is dominated by men and often acts as a woman-free "safe space," one of the terms used to describe it is the *manosphere*, a play on the word *blogosphere*. Those who write about digital culture have used this designation in the past five years or so to signify a loose network of online fora, websites, and social media accounts connected by little other than a shared hatred of feminism. Some of the men who write for these sites self-identify as part of the manosphere, and the term is widely used on the popular blog *Return of Kings*, but others within the community use the term only with irony (and, typically, quotation marks). The term *manosphere* encompasses several of the subgroups mentioned below that focus primarily on gender and sex, and my focus throughout the book will be on the men in these communities. Other groups mentioned are more strongly motivated by concerns of race and ethnicity, and so the term *manosphere* may not apply to them, even though these groups are still vocally antifeminist and deeply concerned with policing female—particularly white female—behavior.

The term *men's rights activists* (*MRAs*) is sometimes used synecdochically to describe all the members of the manosphere, but that usage is not quite accurate. Men's rights activists are only one subset of a highly fragmented community whose factions can be extremely hostile to each other. Paul Elam, a sixty-year-old former substance-abuse counselor and one of the most visible members of the community, has denied that the manosphere as a collective group exists at all: "The very expression, man-o-sphere, implicitly paints an image of connectivity; of shared purpose and identity. Aside from distaste for feminism, which anyone capable of critical thought will share, there is no real or abiding connection; no universality or even commonality, and that lacking manifests in how we tear ourselves, and each other, down, and always have."[9] The deliberate understatement "distaste for feminism" masks the vehemently antifeminist ideology that unites the manosphere's hundreds of thousands of members, but Elam's assessment of the division within the community is accurate.

The primary goal of men's rights activists is the elimination of laws and social norms that they see as fundamentally oppressive to men. These include divorce, child support, and custody laws; routine male circumcision, which they believe is genital mutilation; and the extension of default credibility to women who claim to have been sexually assaulted. Because dismantling oppressive social norms is also a feminist concern, and because these men tend to use slightly less inflammatory rhetoric than other Red Pill subgroups, they have become the most mainstream segment of the community. Their preferred term for themselves is actually not *men's rights activists*, a term often used in a derogatory manner by critics, but rather *men's human rights advocates (MHRAs)*, and they think of themselves as humanists or egalitarians. Their main internet hub is the website *A Voice for Men*, founded by Elam. Elam's experiences with multiple divorces and child support payments for a daughter whose paternity he doubted opened his eyes to how unfairly fathers are supposedly treated. The news website *Vox* wrote of him, "If men's rights activism has a Gloria Steinem, a kind of central activist figurehead, it is Paul Elam."[10] The comparison between Elam and Steinem hints at a larger similarity between feminism and "meninism": like feminists, the men of the manosphere identify structural modes of gender-based discrimination, but their causal explanations differ from feminist interpretations. For example, they see custody laws as overwhelmingly unfair to men because we live in a gynocentric society, while feminists see custody laws as reflecting a deeper problem of gender normativity and biological essentialism that forces women to perform the overwhelming majority of childcare duties.[11]

The men's human rights movement (MHRM) is more inclusive than the rest of the manosphere. Many other factions of the manosphere are male-only spaces, but MHRAs embrace and provide a platform for anyone who supports the mission of identifying and combating misandry, including women and gay men.[12] Women who support the movement are sometimes known within the community as *honey badgers* or *feMRAs*, and they produce a significant portion of the content on A Voice for Men.[13] As far as racial diversity is concerned, in 2015 A Voice for Men did a reader survey and found that 76 percent of respondents identified their race as Caucasian—lower than the manosphere as a whole, but still well above the national average.[14]

The factions within the Red Pill are often hostile to each other, but the animosity between the MHRM and the pickup artist (PUA) community is particularly acute. Pickup artists focus their energy on perfecting techniques for seducing women. To them, the art of seduction is about far more than knowing a few good opening lines. They believe the pickup artist understands both the true nature of women and how they are conditioned to act by society, and that he can exploit that knowledge to make himself attractive to them. This quality is known as *having game*. Members of the men's rights movement see pickup artists as participating in and contributing to gynocentrism; by placing so much value on women as sex objects, they inadvertently afford women power over men. Pickup artists, meanwhile, believe that sexual success is a key element of being a true alpha male, and they believe those in the men's rights movement channel their sexual frustration into social activism because they are unable to convince women to sleep with them.

The seduction community's online presence is spread over several individual blogs run by successful pickup artists with dedicated followers and large internet fora for sharing tips and posting "field reports" and "lay reports," such as MPUAforum.com, which has over 170,000 members, and the subreddit r / seduction, which has over 240,000 members. Pickup artistry gained widespread infamy after the 2005 publication of Neil Strauss's book *The Game: Penetrating the Secret Society of Pickup Artists*. But Strauss's image of how pickup artists communicate with each other—experienced masters of seduction giving in-person seminars to eager followers, groups of pickup artists living together as roommates in rented mansions with names such as "Project Hollywood"—is outdated. The game community is almost entirely virtual now. Many broad-interest Red Pill sites dedicate a significant portion of their content to articles about game advice. One such site is the blog *Return of Kings*, founded and published by Daryush "Roosh V" Valizadeh, which publishes not only game advice but also weightlifting tips, book reviews, and antifeminist screeds—all the information needed to become a true "alpha male."

Although Valizadeh first gained a following as a pickup artist with his blog *DC Bachelor*, he is currently refashioning himself as a theorist of masculinity. He calls his new ideology *neomasculinity*, encompassing values not only of pickup artistry but also of traditional gender roles, male self-improvement, and libertarianism. He introduces his new philosophy in the post "What Is Neomasculinity?" on his personal blog, *Roosh V*, and defines the term:

> Neomasculinity is a new term that uses old ways of helping men live in a virtuous manner while catering to the masculine side of their biological nature. It gives men the practical tools to receive the benefits possible with a male existence while living in natural harmony with women and improving the sustainability and value of their societies. It also provides a man with powerful mental defenses to aid his navigation through a world that wants to reduce him to a zombified consumer who serves at the altar of the corporate state.[15]

"Old ways of helping men" and "powerful mental defenses" are subtle references to the influence of ancient Greek and Roman philosophy, particularly Stoicism, on Valizadeh's thinking—a topic I will return to in the next chapter.

The third and final internal faction of the manosphere is Men Going Their Own Way (MGTOW). Men Going Their Own Way aim to live their lives free of female influence and define manhood completely on their own terms. Over time, this aim has evolved significantly: while it initially started as a movement for self-reliant masculinity in harmony with traditional gender roles, it now advocates for living entirely separate from women and engaging in "marriage strikes."[16] Men Going Their Own Way bemoan the tendency of women to spend their most attractive years with unreliable alpha males—often embodied by the generic characters Chad Thundercock and his black counterpart Tyrone—before they "hit the wall" in their late twenties and their attractiveness begins a steady decline, at which point they become more willing to settle for a beta male.[17] These men believe it is better to opt out of this rigged system.

The community of Men Going Their Own Way is smaller than the men's human rights movement and seduction communities—as of this writing, the site mgtow.com has over twenty-five thousand members and the subreddit r / mgtow has over thirty-five thousand subscribers—and its members disagree on many points with other members of the manosphere. Men Going Their Own Way were in the past almost uniformly libertarian, and their distaste for "big government" led to a schism with the men's human rights movement, many members of which are theoretically interested in activism in the form of lobbying for changes in custody and divorce law. Early Men Going Their Own Way are dismissive of members of the movement as it exists now, and pickup artists are even more derisive, calling the movement "Virgins Going Their Own Way."

In spite of the conflict between pickup artists and Men Going Their Own Way over their differing approaches to women, both groups have begun to merge with the so-called Alternative Right or

Alt-Right, a neoreactionary white nationalist group that began gaining prominence in 2015 and has been steadily growing since.[18] The Alt-Right came into the public spotlight in 2015 and 2016 for its outsized role in the online discourse concerning the 2016 presidential election. After starting a "meme war" in support of Donald Trump and harassing liberal (and particularly Jewish) journalists, the Alt-Right was denounced by Democratic nominee Hillary Clinton in an August 2016 speech in Las Vegas.

The Alt-Right comprises several competing factions, including outright neo-Nazis such as Andrew Anglin, publisher of *The Daily Stormer*. Other less blatantly anti-Semitic members of the Alt-Right, including Richard Spencer, the president of the white nationalist think tank the National Policy Institute, consider themselves identitarians and advocate for a white ethno-state. Spencer gained widespread infamy after giving President Trump a Nazi salute in January 2017. The identitarian subgroup of the Alt-Right also includes Nathan Damigo, founder of the white nationalist organization Identity Evropa, which is famous primarily for putting up posters on college campuses. But arguably the largest faction within the Alt-Right is the group of anonymous trolls, or "shitposters," who dedicate their time to spreading abuse online, much of it sexist and racist.

The Alt-Right might appear to be an outlier alongside the manosphere, since its primary focus is racial rather than gendered. However, many members of the Alt-Right are also either pickup artists or Men Going Their Own Way: James "Roissy" Weidmann, who runs the popular game blog *Chateau Heartiste*, is an open white nationalist, as is Theodore "Vox Day" Beale, a science-fiction writer who runs a blog called *Vox Popoli* (*sic*), dedicated to politics and science fiction, and another called *Alpha Game*, dedicated to pickup. Furthermore, the policing of white female sexuality is a major concern for all alt-right men, who rail against the societal ills of "race-mixing."[19] Their vision of an ideal future for the white race cannot materialize if white women do not bear the children of white men. They have therefore adopted the term *white genocide* to describe the rise in interracial relationships. Not all interracial relationships are equally offensive to them, however: Anglin once wrote, "I'm not really triggered by miscegenation if the mixer is a male, because it doesn't really make any difference, biologically. With a woman, there is a lot of anger, because it's OUR WOMB—that's right, it doesn't belong to her, it belongs to the males in her society—that is being used to produce an enemy soldier. Our race isn't going to run out of sperm any time soon, so when it's a man doing it, it's just sort of like 'lolwut?'"[20]

The process by which men find the Red Pill accounts for both the high degree of overlap with these groups and the tensions between them. Although the fora I have mentioned continue to experience gradual growth, certain events have led to large-scale Red Pill moments that attracted thousands of new members. One such event was the GamerGate movement, which began in August 2014 as a backlash against the rise of progressivism and feminism in videogame culture. In theory, the movement is concerned with ethics in gaming journalism and the perceived tendency on the part of journalists to give disproportionately more positive reviews to games that cater to "politically correct" tastes. In reality, GamerGate is, as the *Washington Post* described it, a "freewheeling catastrophe / social movement / misdirected lynch mob."[21] As that movement began to lose momentum, many public faces associated with GamerGate—including Beale, the former *Breitbart* technology editor Milo Yiannopoulos, and the writer Mike Cernovich—shifted their focus from videogame politics to national politics and became outspoken supporters of Donald Trump, whose nomination provided another large-scale Red Pill moment. Some of these writers, however, seem leery of explicitly allying themselves with outright white supremacists and anti-Semites such as Andrew Anglin. Instead, they have opted for a softer version of alt-right politics—sometimes called *alt-light*—that focuses on preserving the culture and values associated with "Western civilization" against threats both external (such as Muslim refugees) and internal (such as progressive intellectuals).[22]

These disparate units of the Red Pill are, however, united by a common enemy: *social justice warriors* (SJWs). This blanket term covers anyone who shows support for the rights of women, people of color, people of diverse sexualities and genders, and those with disabilities. In an article from October 2014, "What Is a Social Justice Warrior (SJW)?" Valizadeh uses this definition: "Social justice warriors believe in an extreme left-wing ideology that combines feminism, progressivism, and political correctness into a totalitarian system that attempts to censor speech and promote fringe lifestyles while actively discriminating against men, particularly white men."[23] I will return later in this chapter to the tactics the Red Pill uses to combat the SJW threat.

Classics beyond the Red Pill

Now that I have established who the men in the Red Pill *are*, the question remains: why are they so interested in the classical world? In fact, their interest in the Classics is not as surprising as it may at first seem; it is only the latest development in the millennia-long use of classical antiquity to promote reactionary ideologies.

Scholars use the term *the classical tradition* to describe the long history of post-classical societies positioning themselves as the heirs of ancient Greece and Rome. Examples of this self-fashioning can be found everywhere in United States' history, from the references to ancient government in the Federalist Papers, to the Doric columns on our national monuments, to the rewriting of Homer's *Odyssey* into the 1930s South in *O Brother, Where Art Thou?* Even the practice of throwing toga parties at fraternities serves as a convoluted, ahistorical reimagining of the relationship between the modern university and Plato's Academy.

Where the classical tradition thrives, the greatest enthusiasm for the ancient world is usually found among social elites. The similarity between the term *Classics* and the word *class* to denote social status is not accidental. From the first Roman census, believed to have been conducted by the sixth king of Rome, Servius Tullius, we learn that in the sixth century BCE, Rome's population was split into six groups, each called a *classis*. British classicist Edith Hall writes,

> It is to the legendary first census that there must also be traced the origins of the term *Classics*. In Servius' scheme, the men in the top of his six classes—the men with the most money and property—were called the *classici*. The top men were "Classics," and this is why, by the time of the late second-century AD Roman miscellanist Aulus Gellius, by metaphorical extension the top authors could be called "Classic Authors," *scriptores classici*, to distinguish them from inferior or metaphorically "proletarian" authors, *scriptores proletarii*.[24]

Knowledge of the *scriptores classici* was then typically reserved for the *classici* census group. Being well-versed in the Classics is almost by definition a signifier of high social class. Hall recounts anecdotes of middle-class British men during the eighteenth and nineteenth centuries painstakingly educating themselves in Greek and Latin in order to find greater professional success.[25] Knowledge of the Classics is used in these narratives to reinforce existing hierarchies. The potential for upward social mobility through classical education is not truly democratic or revolutionary; it merely continues the trend of excluding those who are *not* knowledgeable about Greece and Rome.

From this perspective, the far-right attraction to the classical tradition is logical. The most active members of the Red Pill community are usually white men who believe that white men are being oppressed. When they look back to the dead white men of the ancient world as the sources of ultimate wisdom, the takeaway is that white men *are* better. They gave us the beginnings of our philosophy, culture, and art. They—that is, we, their white male descendants—*deserve* to be in charge. Andrew Anglin articulates this reactionary viewpoint in "A Normie's Guide to the Alt-Right":

> *Endorsement of White History*
> The Alt-Right celebrates the greatness of our ancestors and the glory of our historical achievements. Rejecting revisionist arguments by modern social scientists which portray Whites as having wrought evil on the planet, we view Whites as the creators and maintainers of Western civilization.[26]

Immediately underneath this declaration is a photograph of the Colosseum in Rome.

Conservatives have long invoked ancient history to idealize and argue for a lost past. In *Trojan Horses: Saving the Classics from Conservatives* (2001), classicist Page duBois deconstructed some of the most egregious examples of this trend from the late twentieth century: "In such settings, the culture of ancient Greece is once again, as before in American culture, reduced and manipulated to persuade readers and viewers of the immemorial truth of the most illiberal of political opinions."[27] She continues, "Theirs is a selective and impoverished version of ancient culture, one that for the most part erases historical difference and, looking through the window of history, finds the Greeks motionless as in a diorama, caught in a tableau vivant exemplifying the moral virtues of conservative twenty-first-century America. Conservatives fetishize a particular stereotype of the Greeks by fixing and repeating it, and they distort history for their political purposes."[28]

DuBois's book is important, because she recognizes that these writers use the Classics as a battleground. Interpretation of ancient Greek and Roman material becomes a medium for promoting contemporary political agendas. The ancient world represents a time that was less progressive, both technologically and socially, and, while conservatives tend to applaud technological progressiveness, they also tend to bemoan social progressiveness. The same tactic, in fact, is used by the men studied in this book, although not always at the same level of intellectual sophistication.

The men of the Red Pill are particularly attracted to the ancient world because they see in it a reflection of their own reactionary gender politics. Much of the literature that has survived from the ancient Mediterranean world can be adapted easily to their ideologies, and there is a deep well of ancient misogyny to draw on. Generations of classicists, especially feminist classicists, have had to confront these problems when reading the poetry, drama, epics, histories, and philosophies of their chosen field. Aristotle's theory of natural slavery and the inferiority of women is considered one of his least philosophically sound ideas by classicists, but it lends itself well to Red Pill ideology—and, indeed, in 2015, *Return of Kings* published an article titled "Feminism Comes Full Circle into Embracing Aristotle's 'Natural Slavery.'"[29] In such articles, Red Pill writers use the ancient world for its instructive exemplarity. These articles are worth looking at more closely, because their analyses of openly misogynistic ancient texts play a role in defining what the ancient world *means* in the context of the Red Pill: namely, that all women, over all historical periods, share the same negative characteristics.

In response to discussions about the essential nature of women on Red Pill sites, one acronym that frequently appears is NAWALT, which stands for "not all women are like that." NAWALT is a signifier for men who have not fully swallowed the Red Pill and insist that there exist in the world exceptional, extraordinary women who do not share the negative traits the Red Pill community usually attributes to *all* women.

The discourse in the Red Pill community concerning the phrase *not all women are like that* resembles the feminist critique of the phrase *not all men*, which entered the vernacular of feminists active online in mid-2014.[30] *Not all men* signifies a common male response to discussions about systemic sexism in which a man insists that he is not personally sexist and therefore is not part of the problem. One word often used in this context is *centering*—that is, redefining the central issue or perspective in a discussion. *Not all men* centers the male perspective on misogyny, privileging the interlocutor's personal feeling that he is being misrepresented and his virtue is being improperly appreciated over any discussion of widespread sexism. It demands that the man be praised for not contributing to the problem without his even fully acknowledging that the problem is real.

Not all women are like that and *not all men* both acknowledge the existence of sexist discourse, although the former explicitly validates sexism while the later ostensibly portrays sympathy for feminism. But both rhetorical strategies use an appeal to exceptionalism to avoid explicit participation in sexist structures. And just as feminists have critiqued the concept of *not all men*, the men of the Red Pill prefer the opposite of *not all women are like that*: *all women are like that* (AWALT), a programmatic declaration of universal female behavioral characteristics that underlies much of their ideology. Phrases such as *all women are like that* posit a sameness among women that erases specificity. Temporal continuity from the ancient world to today allows for a natural expansion of that idea, namely that all women are—and have always been—like that, because female nature is essentially the same between all women, in all periods of time, from ancient Greece and Rome to today.

———————

Misogyny appears early in Greek literature. Near its beginnings, we find the idea that women are parasites and men would be much better off if the gods had granted them the ability to reproduce asexually. This is the poeticized view of Hesiod, whose *oeuvre* is usually dated to the eighth century BCE. Two surviving poems are attributed to him: *Theogony*, about how the Olympian gods came to rule over the cosmos, and *Works and Days*, which is addressed to his lazy and shiftless brother Perses, with advice for how to become a properly self-subsistent farmer. In both of these texts, the speaker asserts that men would be better off without women. In *Works and Days*, he explains why it is necessary for men to labor for food: once upon a time, the world was tremendously fertile, and almost no labor was required to survive. An entire year's worth of work could be done in a single week (*Works and Days* 43–44). Because the Titan Prometheus stole fire for humans, Zeus, the king of the Olympian gods, retaliated by creating the first woman, Pandora, a "beautiful evil" (*kalon kakon*) for mankind (*Theogony* 585). He writes of Pandora that

> She founded the tribe of females,
> The destructive tribe of women
> Who live amongst mortal men and cause great pain,
> Companions in wealth but not in poverty.
> That is how Zeus, the loud-thundering one, made women:
> As an evil for mortal men, a troublesome partner.
> And he added an additional evil to offset the good:

Whoever flees marriage and the mischief of women
And does not wish to marry, that man arrives at destructive
    old age
Without any children to take care of him. He is prosperous
While he lives, but when he is gone, his relatives
Divide up his possessions. (*Theogony* 591–607)[31]

Considering the tone and content of this passage, it is no surprise that offhand references to Hesiod show up on occasion throughout the manosphere to prove that misogyny has ancient roots. In a 2016 thread on r/theredpill with the title "To Marry or Not to Marry? An Ancient Perspective," the poster uses an excerpt from *Works and Days* to show that marriage is a "damned if you do, damned if you don't" kind of arrangement. He prefaces his claim by saying, "I am a classicist by training, PhD the whole nine yards. The Greeks and Romans were Red Pill in the extreme."[32]

The idea that women are useless parasites lives on in later Greek literature, appearing again in the lyric poet Semonides, from the island Amorgos in the Greek Cyclades, who reportedly flourished in the seventh century BCE. In his most famous fragment, the seventh, he lists ten different types of women, compares them to animals, and explains why nine of them are awful. The horse-woman spends too much time grooming herself; the sea-woman is unpredictable; the pig-woman is filthy and obese. And then there is the donkey-woman: "She's hard-headed and obstinate, and will barely do her work, even after you threaten and force her. And even then, she won't finish. She stays at home and eats all day and all night. And she's just as greedy when it comes to sex: she'll take any man she sees" (43–49). Only the bee-woman, who industriously tends to her home, will make a good and useful wife—but the narrator warns that even wives who initially seem virtuous are likely hiding their vices, turning their husbands into cuckolds and laughingstocks (83–95, 108–114).

Even when one allows for the extreme cultural differences between the United States in the twenty-first century CE and archaic Greece, this fragment is a remarkably misogynistic text. In fact, that is probably its point. The idea that "Zeus made marriage the hardest struggle for men, an unbreakable bond" seems to echo Hesiod's depiction of marriage and women as intentionally created as a punishment for men (Semonides, seventh fragment 115–116). Both texts ostensibly blame Zeus for bringing down misfortune on men in the form of sexual reproduction, but despite this framing, criticism of Zeus is clearly secondary to antiwoman vitriol.

The poems of Hesiod and Semonides show the development over time of an established rhetoric of misogyny, a pattern of familiar language for complaining about wives—without ever considering how wives feel about being objects of exchange between their husbands and fathers.[33] This antimarriage sentiment finds its clearest echoes in the Men Going Their Own Way faction of the manosphere, and some in this movement have taken notice. The blog *No Ma'am*, one of the first sites dedicated to the community, having published a "MGTOW Manifesto" in 2001, included Semonides in a 2010 list of "Great Zingers" against women.

Not all Greeks thought that women were useless, but many believed that women were inferior and needed to be trained into utility. There are several texts that take the form of advice for men on how to control female behavior, not unlike the relationship advice traded in articles on pickup artist websites. One key text in this genre is Xenophon's *Oeconomicus*. Xenophon was a contemporary of Plato, and this text is a dialogue between Socrates and another individual, a conversational format similar to that in many of Plato's works. In *Oeconomicus*, Socrates and a man named Critobulus discuss wealth and the proper running of a household, and Socrates tells the story of a man he knows named Ischomachus, whose wife, according to Socrates, is remarkably skilled at tending to their home. Her name is never given.[34]

Xenophon's depiction of marital harmony fits perfectly with the manosphere's vision of ideal "pair bonding."[35] When Socrates asks Ischomachus how his wife came to be such a good housekeeper, Ischomachus recalls several conversations in which he led her, by way of the Socratic question-and-answer method, to understand the proper roles of the husband and wife. First, he explains to her that the complementary strengths of men and women make men better suited to work outside the home and women better suited to work within it (*Oec.* 8). Then he convinces her to see herself as the general of the house's army (*Oec.* 9). He also convinces her that no wife is more attractive to her husband than a good housekeeper: "I said that mixing flour and kneading dough were good exercise, and so were shaking and folding clothes and linens. I said that this kind of exercise would give her a good appetite, good health, and it would bring a natural flush to her cheeks" (*Oec.* 10.11).[36]

Citations from misogynistic Roman literature also appear on Red Pill sites. Juvenal (Decimus Iunius Iuvenalis) was a Roman satirist who lived in the late first and early second centuries CE. Although Juvenal rails in his satires against Roman men, Roman society, and the city of Rome itself, Red Pill sites tend to focus on his sixth satire, a 650-line manifesto arguing that men should not marry because any woman will eventually make her husband miserable.[37] It offers such advice as

> Even if she is beautiful, honorable, rich, fertile, and has
> Ancient ancestors in her porticoes; even if she is more
>   chaste
> Than a Sabine woman with flowing hair who ended a war,
> A rare bird, like a black swan on Earth:
> Who would be able to stand such an exemplary wife? I'd
>   rather
> Have a prostitute than you, Cornelia, mother of the
>   Gracchi,

If along with your multitude of virtues you bring
A proud eyebrow and triumphs as part of your dowry.
(162–169)

The manosphere has noticed how congenial this description is to their own views; on r/theredpill, one redditor writes that Juvenal "was apparently an ancient TRPer and was writing to convince guys not to marry the women of his time because of all the same vices we see today. From sleeping around (riding the cock carousel) to hamstering to hypergamy, it was all there."[38] One writer on *Return of Kings*, analyzing the same text in the article "There Is Little Difference between Women throughout History," concludes, "Even in an era when the goddamned patriarchy supposedly ruled with an iron fist, an age that knew no concept of effective birth control, a time when being a single mother would assuredly leave an ordinary woman destitute and forever burdened; women were still sluts."[39]

This survey of misogyny in classical literature is far from complete; in fact, it only addresses some of the most egregious and explicit instances of woman-hating, glossing over the multitude of texts that erase women, silence them, or condescend to them. Some of the men in the manosphere are aware of these openly and loudly misogynistic texts, and occasionally they write analyses of them. These analyses showcase the learning of the writers and elicit a chorus of remarks in the comments section about how men have always known what they are dealing with when it comes to women.

Such articles reinforce the comfortable idea that the ancient Greeks and Romans had the same ideas about gender that the Red Pill community holds. The study of the use and discourse of masculinity in spaces such as the manosphere—something I have seen termed *meninism* or *masculism*—is a relatively nascent field, lacking context, depth, and precision.[40] Instead, it relies on the arrogance that its practitioners are the inheritors of the classical tradition. This idea is supported by the fact that many ancient texts *do* exhibit concerns that remain relevant to the manosphere today, from emotional control to seduction to false rape allegations. So how could a woman possibly understand Aristotle as well as a man could, when Aristotle himself claimed that women were intellectually inferior?

Dead White Men and Angry White Men

Of course, although there is much misogyny to be found in the literature and history of classical antiquity, not all dead white men were like that. Some even had what might generously be described as protofeminist leanings. But there is no denying that producing feminist readings and uses of the Classics can be a bit like trying to use a normal pair of scissors when you are left-handed: they were designed with somebody else in mind. And some people—a few in the Red Pill, a few in academia—even believe that feminist Classics is impossible and that feminism and interest in the ancient world are natural enemies.

In a 2016 article on *Breitbart*, "An Establishment Conservative's Guide to the Alt-Right," Allum Bokhari and Milo Yiannopoulos attempt to describe the forces that attracted various groups to the alt-right movement. They argue that one of these groups, the "natural conservatives," are motivated by the devaluation of Western culture, especially on college campuses:

> These attempts to scrub western history of its great figures are particularly galling to the alt-right, who in addition to the preservation of western culture, care deeply about heroes and heroic virtues.
>
> This follows decades in which left-wingers on campus sought to remove the study of "dead white males" from the focus of western history and literature curricula. An establishment conservative might be mildly irked by such behaviour as they switch between the State of the Union and

the business channels, but to a natural conservative, such cultural vandalism may just be their highest priority.

Bokhari and Yiannopoulos are exaggerating, of course—curricula may no longer exclusively comprise dead white men, but that demographic is still far from underrepresented—however, the trend they identify is real. There are some progressives who would like to entirely destabilize the canon, including the Greek and Roman classics. Contrary to what the *Breitbart* writers suggest, however, those with such views represent a small minority, albeit a vocal one.[41]

One highly publicized incident from 2015 concerning the works of Ovid seems to illustrate this trend. Ovid—Publius Ovidius Naso—was born in Sulmo, a town to the east of Rome, in 43 BCE. His poetry was composed during the time of Rome's first emperor, Augustus. Although his most famous work is the epic poem *Metamorphoses*, his *oeuvre* also comprises love poetry, an epic poem about the Roman calendar, a lost tragedy *Medea*, and works written after he was exiled by Augustus to Tomis, on the shore of the Black Sea. I will return to Ovid and his *Ars Amatoria* in Chapter 3, but much of the modern-day controversy around Ovid centers especially on his magnum opus, *Metamorphoses*, a poem in fifteen books about gods and men who changed from one form into another. One of the traumatic events that can lead a character to a change in form is rape, and the text contains many rape narratives—perhaps more than fifty, including nineteen told at length.[42]

A 2015 op-ed in the *Columbia Spectator* by four members of Columbia's Multicultural Affairs Advisory Board describes how reading Ovid's *Metamorphoses*, a text assigned as part of Columbia's Core Curriculum, triggered post-traumatic stress disorder symptoms in a student who was a survivor of sexual assault:

During the week spent on Ovid's "Metamorphoses," the class was instructed to read the myths of Persephone and

Daphne, both of which include vivid depictions of rape and sexual assault. As a survivor of sexual assault, the student described being triggered while reading such detailed accounts of rape throughout the work. However, the student said her professor focused on the beauty of the language and the splendor of the imagery when lecturing on the text. As a result, the student completely disengaged from the class discussion as a means of self-preservation. She did not feel safe in the class. When she approached her professor after class, the student said she was essentially dismissed, and her concerns were ignored.

Ovid's "Metamorphoses" is a fixture of Lit Hum [Literature Humanities], but like so many texts in the Western canon, it contains triggering and offensive material that marginalizes student identities in the classroom. These texts, wrought with histories and narratives of exclusion and oppression, can be difficult to read and discuss as a survivor, a person of color, or a student from a low-income background.[43]

The writers of the op-ed asked that the university institute what are known as *trigger warnings* for texts such as Ovid's.[44] This request was not granted, although the *Metamorphoses* has since been replaced in the Curriculum Core with selections from Ovid's *Heroides*, a series of poetic letters written by the women of myth to the heroes who abandoned them.

The *Columbia Spectator* op-ed was met with bipartisan scorn in outlets from *The Wall Street Journal* to *Salon*.[45] The conversation even continued on Red Pill fora. On a GamerGate subreddit, a user shared a post about the controversy under the title "Triggered SJWs Have Successfully Gotten Ovid's 'Metamorphoses' Removed from the Syllabus for a Required Core Course at Colombia [*sic*] University." The post complained that "Social Justice Warriors have reduced universities to places that pander to the lowest common denominator. The most pathetic, whining, imbecilic losers are the ones who are in charge."[46] *Breitbart*, the far-right news website beloved by the Red Pill community, also responded to the incident with a blog post, this one titled "Campus Special Snowflakes Melt upon Contact with Greek Mythology."[47]

These responses, like the ones in the mainstream media, imply that Ovid's poem deserves a place in university classrooms not for its content, but because it is old and canonical. It is this kind of thinking that makes ancient Greece and Rome so attractive to the Red Pill; the Western Classics are valued more for being Western and classical than they are for their content. Crucially, Red Pill writers misunderstand the nature of the students' request: Ovid's works remain on the syllabi at Columbia, and the author's place in the canon has not really been challenged. Nevertheless, the incident was treated by the Far Right as an assault on dead white men.

The Columbia op-ed was written by a group of students, but some senior feminist scholars do argue that studying the Classics, along with the rest of the Western canon, is work that necessarily reifies patriarchal and racist hierarchies. Feminist scholar Sue-Ellen Case makes this argument about Greek tragedy and comedy in her essay "Classic Drag: The Greek Creation of Female Parts." Case concludes, "Feminist practitioners and scholars [of Greek drama] may decide that such plays do not belong in the canon—that they are not central to the study and practice of theatre."[48]

This specter of the elimination of classical texts has been very useful for the Red Pill community. In a volume of essays titled *Thirty Seven*, manosphere writer Quintus Curtius ("QC")—formerly a popular columnist on the manosphere blog *Return of Kings*—imagines a dystopian future in which feminists have rewritten the canon and erased the Classics: "One can even imagine a future where classical knowledge will be driven underground, purged from schools, or bowdlerized, as not being in tune with modern feminism and political correctness. The degradation of humanistic learning has come as a direct result of the feminization of American society. We cannot permit this to happen. The commissars of modern culture don't want you to know too much about history, or about how things were like in previous eras."[49] Women and feminists, it seems, are the enemies of the future of the ancient.

Valizadeh, perhaps the most famous misogynist in the manosphere, takes this vision a step further in his 2014 article "What Is a Social Justice Warrior (SJW)?" in which he theorizes about the complete moral vacuum that would exist in a society with no classical literature:

> Even if Aristotle, Seneca, Marcus Aurelius, Thomas Aquinas, or Henry David Thoreau had valuable wisdom that continues to help how millions of people live today, the information derived from their work must be completely discarded since they were white men. Since white men were at the forefront of advancing humanity for the past several centuries, especially after the decline of the Egyptian, Persian, Mongol, and Ottoman empires, this precludes the bulk of moral guidance that we can use to determine right and wrong. SJW's invent their own moral code but it is often based on what they are upset about in the present moment. It does not serve as a guide for more than a month or two, suggesting that their book of code would have to be written in pencil.[50]

It takes considerable rhetorical maneuvering to argue that texts that have been read for thousand years are suddenly in immense imminent danger and require staunch defense by Red Pill men. But Quintus Curtius and Valizadeh are also echoing a strain of thought that exists within the discipline of Classics, perhaps best exemplified by the Victor Davis Hanson and John Heath polemic *Who*

Killed Homer? (1998). Even earlier, in the 1987 book *The Closing of the American Mind*, Allan Bloom had written that "the latest enemy of the vitality of classics texts is feminism."[51] His use of the word *vitality* suggests that ancient texts are somehow not as "alive" for feminists as they are for misogynists.

Bernard Knox, an eminent British scholar of Greek tragedy, expressed a milder version of this view in the essay "The Oldest Dead White European Males." First published in *The New Republic* in 1992, then republished in a slim volume of his essays by the same name in 1993, the essay argues against what he sees as a pernicious trend in Classics: modern classicists who study anthropology, feminism, and slavery are fixated on how *strange* and *different* the Greeks were compared to us. Knox believes that the Greeks have been "othered" by modern scholars, made into an alien species. The classification of the Greeks as DWEM, Dead White European Males, marks them as entirely separate from us.[52]

Knox writes:

> Their assignment to the DWEM category is one of the accomplishments of modern multicultural and feminist criticism; and it is a declaration of their irrelevance. But previous ages spoke of them in very different terms. "We are all Greeks," wrote Shelley in 1822, "our laws, our literature, our religion, our arts, have their roots in Greece." If the Victorian vision of Greece could be summed up in the slogan GREEKS 'R' US, the modern critics could retort that GREEKS 'R' THEM, or more pointedly, GREEKS 'R' DWEM.[53]

He concludes, "Though we can no longer say, with Shelley, that we are all Greeks, nor can we claim, as the Victorians might have claimed, that GREEKS 'R' US, we must always acknowledge how greatly, how deeply, how irrevocably, we remain in their debt."

This argument, much like those of the Red Pill writers I mentioned earlier, is a straw man: nobody denies that our society owes a debt to the Greeks. The question is how that debt should be treated. Should we romanticize that debt, as Shelley did, and as one can find men on Red Pill websites doing today? Or is it a more complicated and problematic legacy? And if the Greeks really are "other" and not as similar to us as the Victorians thought, would that difference make them less worth studying?

Knox also argues that the ancient Greeks are not as different from Americans as we would like to think, because we share some of their notable failings:

> Indeed, when we think of the two great flaws in Athenian democracy that recent scholarship has explored and emphasized, we ought to remember not only that slavery and male dominance were characteristic of all ancient societies, but also that we, of all people, have no right to cast the first stone. Pericles' proud claim for Athenian democracy— power in the hands of the people, equality before the law— makes no mention of the slaves, but our Declaration of Independence, according to which "all men are created free and equal," does not mention them either, although the man who drafted it and many of those who signed it were owners of African slaves.

Both the men of the Red Pill and those who are interested in social justice are perfectly aware that slavery and male dominance are written into the foundation of and still exist in our country in a very real sense. The carceral state has maintained a status quo in which a large percentage of the black adult male population is still unfree— sometimes called *the New Jim Crow*—and white men still dominate government and industry on nearly all levels.[54] We do not need to look back to the Founding Fathers to find the flaws in our country that Knox admits existed in ancient Greece. Indeed, the Red Pill community believes these very aspects of early American society made America great in the past and must be resurrected to make America great again.

These similarities between our society and ancient Greece—and Rome—are beneficial to the men in Red Pill communities, because they make it all the easier for them to see a reflection of their own values in the ancient world.

The Red Pill Toolbox

Feminists on the internet sometimes attempt to engage directly with Red Pill men on these issues, but these encounters often end in frustration. Arguing with them is nearly impossible, because these men have a set of established rhetorical tricks that they use, consciously or unconsciously, to misdirect and derail opposing views. Since this book is dedicated to exposing how Red Pill rhetoric works, these tricks deserve some explication.

The first of these is based on a concept used often by pickup artists called *frame theory*. Frame is a subject's perspective on the world, themselves, and any situations they encounter. A pickup artist might have the frame that he is an important and valued man in his community, and therefore any woman would be fortunate to sleep with him. His task then becomes *maintaining frame* (or *frame control*) in interactions with beautiful women and, ideally, influencing them so their frames become assimilated to his.

There is another, more commonly used name for this technique: *gaslighting*. Gaslighting is an emotional abuse tactic in which one person tries to convince another that they cannot trust their memories and perception. This technique gets its name from the 1938 play *Gas Light* (performed in the United States as *Angel Street*, and released as the film *Gaslight* in 1944) by Patrick Hamilton, in which a husband tries to convince his wife that she is insane.

Frame control is essentially a form of gaslighting: the goal is to convince someone that your version of reality is more accurate than theirs.

When feminists attempt to debate their differences rationally with pickup artists, Men Going Their Own Way, men's human rights advocates, or members of the Alt-Right, they are met with gaslighting. Your frame is wrong and their frame is right. The result is that people with destructively unbalanced readings of literature and current events, readings that are impervious to reasoned critique, have an audience of hundreds of thousands of angry men. Many of them, as I will argue in the next chapter, also claim to be Stoic and believe that they are in control of their rage, while those who disagree with them are angry / rabid feminists.

As an example of how these men respond to critique, here is the response Theodore "Vox Day" Beale published to a critique of the Alt-Right by writer Cathy Young. Beale, as I mentioned earlier, gained visibility during the GamerGate movement, although he has since shifted his concerns from the internal politics of the science-fiction community to politics on a national scale. In a post on his *Vox Popoli* (*sic*) blog with the title "(((Cathy Young))) Critiques the #AltRight," Beale responded,

> There is no attempt to avoid the "white supremacist" label that (((Ekaterina))) and other opponents of the Alt-Right are so desperate to pin on it. I am an Anglo-Aztec-American Indian who is genetically superior to 99 percent of all blacks on the track and intellectually superior to 99 percent of all whites and Jews, so I'm not inclined towards white supremacy myself.[55]

Between his use of the triple-parentheses echoes—an anti-Semitic dog-whistle used by the Alt-Right as the virtual equivalent of the yellow Jewish star badge used in Nazi Germany—and his declaration of genetic superiority, Beale proves that Young was right: the Alt-Right is a movement powered by racial discrimination. He also preempts potential criticism with his assertion that he is "intellectually superior to 99 percent of all whites and Jews"; other racial groups, one assumes, are so self-evidently intellectually inferior as to not deserve his mention. Since Beale believes himself to be smarter than anybody worth mentioning, his critics must be too unintelligent to understand him properly.

Choosing not to engage with these men directly, however, does not preclude critiquing their views. In fact, such critiques are crucially necessary, because they can untangle the frequent attempts by Red Pill members to derail conversations by intentionally obfuscating the issues at stake. This is the second tool in their rhetorical misdirection toolbox, which I have termed *the appropriative bait-and-switch*. This technique borrows the language of systemic oppression from social justice movements while intentionally introducing confusion about who, precisely, is being oppressed.

The appropriative bait-and-switch is evident in this declaration from the essay "The Misandry Bubble" by Imran Khan, an influential text in the manosphere listed as "essential reading" on the Red Pill subreddit: "Modern misandry masking itself as 'feminism' is, without equal, the most hypocritical ideology in the world today. . . . Men have been killed due to 'feminism.' Children and fathers have been forcibly separated for financial gain via 'feminism.' Slavery has returned to the West via 'feminism.' With all these misandric laws, one can fairly say that misandry is the new Jim Crow." When the men of the manosphere claim that forcing fathers to pay child support is the new Jim Crow (and ignore the fact that *the new Jim Crow* is already a term used to talk about the mass incarceration of black men), they appropriate to disastrous effect a topic that is about race and the legacy of slavery and use it to support an ideology that allows white men to restrict women's reproductive freedom by limiting access to abortion and birth control.

This technique is used far more commonly than one might think. When author Harper Lee died in 2016, the major men's human rights movement website *A Voice for Men* published an article about how "all men are Tom Robinson now," claiming that *To Kill a Mockingbird* was not really about race but about *prejudice*, the real victims of which are the men who suffer through false rape allegations as Tom Robinson did.[56] This article uses a timely issue, the death of a famous author, to appropriate racial concerns in order to further benefit white men. And use of this tactic extends beyond turning racial issues into gender issues. Many articles on Red Pill websites insist that rape culture in this nation is a fabrication, but that true rape culture can be found in the sexual assaults committed by Muslim immigrants (whom they call *rapefugees*). This formulation effectively derails discussion of rape culture and turns the conversation toward encouraging bias against Muslims and immigrants. Liberals who welcome refugees and protest anti-immigrant policy are labeled "cucks" who are subject to both figurative and literal cuckolding: figurative because the interests of another race are being elevated above the white race, and literal because they are inviting Muslim men to come and sleep with white women, whether or not the women consent.

The bond that keeps the Red Pill community together is ideological sublimation: racial to gendered, gendered to racial. This trend is even more obvious in their intentional, explicit blurring of the line between the sexual and the political. A commenter on an article in *The American Conservative* writes, "The connection between the Alt-Right and the various masculinity movements deserves some elaboration. As many have mentioned, the demographics of the Alt-Right are young and male, and the role of young male frustration (sexual or otherwise) cannot be overstated."[57] Sexual frustration becomes political frustration. Pickup artists support Donald Trump.[58] The Alt-Right insults conservative politicians who waste their energy trying to compromise with the Left by calling them *cuckservatives*, a portmanteau of *cuckold* and *conservative*. Political weakness is equated with figurative emasculation. The line between sexual impotence and political impotence is blurred.[59]

Another, related tool is their misuse of the language of scholarly interpretation. One word that comes up frequently on Red Pill websites is *narrative*. The tagline of A Voice for Men is "Changing the Cultural Narrative." Articles on *Return of Kings* have titles that include phrases like *the liberal narrative, the multiculturalist narrative*, and *the establishment's narrative*. But, as so often in the Red Pill, a coined phrase or term takes the place of actual interpretation. What precisely the narrative is, or how narrative as a concept is useful, never gets defined.

The final rhetorical misdirection tool I want to mention here is the false equivalence. This tactic can take several forms, but the most commonly seen form on Red Pill sites is "feminist concern A may be bad, but it is similar to Red Pill concern B that feminists are apathetic about. Therefore, feminists are disingenuous hypocrites." This tactic manifests most frequently in connection with discussions about rape culture. Feminists are labeled hypocrites for fighting against rape culture but not caring sufficiently about false rape allegations, or prison rape, or the use of sexual assault by ISIS, or some other marginally related issue.

All of these tactics will appear throughout this book, both in how Red Pill websites approach contemporary issues and in their analysis of the ancient world.

One could argue that it is anachronistic and unfair to call authors such as Hesiod, Xenophon, and Juvenal misogynistic or sexist. I will return to this topic briefly in the next chapter, when I explore feminism and sexism in the works of Stoic philosophers, but my interest is not in inflicting our own cultural anxieties and concerns on ancient Greece and Rome. Instead, I want to show how that projection is implicit in how Red Pill writers approach the Classics. The groups I described in this chapter—the men's human rights movement, the seduction community, Men Going Their Own Way, the Alt-Right—claim that study of the classic texts of the Western tradition is under attack by progressives, who wish to either throw out the canon or read today's corrosive identity politics into it. But in reality, these groups are the ones reading themselves into classical antiquity. These men see themselves reflected in ancient Greek and Roman writers, and they use references to ancient literature as a site for discursive negotiation of their place in the history of white male culture. They elide the immense differences between our society and classical antiquity to attempt to prove the incontestable value of patriarchy and white supremacy—and to argue for the reinstatement of those oppressive systems of power today.

These Red Pill analyses of ancient texts may seem simplistic and misguided to us. In fact, they are not really producing analyses at all. Their interpretations of the Classics should be approached not as readings of the ancient world, but rather as aspirational representations of the world they wish we inhabited. They idealize a model for gendered behavior that erases much of the social progress that has been achieved in the last two thousand years—and they are using ancient literature to justify it.

2

THE ANGRIEST STOICS

Epictetus, a Stoic thinker and former slave, summarizes one of the central concepts of Stoic philosophy in the beginning of his *Enchiridion* (*Handbook*): the importance of recognizing what is within one's own power and what is not.

> There are things that are within our power and things that are not. The things within our power include opinion, motivation, desire, aversion: in short, whatever is our own action (*hēmetera erga*). Not within our power are our body, possessions, reputation, and political position: in a word, whatever is not our own action. Remember that if you regard . . . that which is not your own as your own, you will be hindered, you will lament, you will be disturbed, you will blame the gods and other men. But if you believe—correctly—that only your own things are your responsibility, and other people's things are not, then nobody will hinder you, you will blame nobody, you will accuse nobody.

Two thousand years later, Daryush "Roosh V" Valizadeh blamed women for turning him to the manosphere in a post called "You Did This to Me":

> You made me a selfish asshole. You rewarded me with sex when I treated you poorly. Your pussy got wetter the less I

respected you. You made me go against my kind nature by being more cocky and arrogant.[1]

Epictetus advises taking responsibility for our perceptions and actions; Valizadeh claims that women forced him to act in a certain manner.

To quote Seneca, another Stoic thinker, "Why do you compare two things that are so dissimilar as to be opposites?" (*De Vita Beata* 7.3; *Quid dissimilia, immo diversa componitis?*). The schools of thought from which Epictetus and Valizadeh originate—Stoicism and Red Pill ideology—seem irreconcilable. I will describe Stoicism in more detail below and explain who Epictetus and Seneca were, but in brief, Stoicism, a philosophical school founded in Athens around 300 BCE by Zeno of Citium, teaches its adherents that nearly everything usually perceived to be harmful (including hunger, sickness, poverty, cruelty, and death) is only harmful if one allows it to be. The only true evil is vice. In contrast, the Red Pill is united by its belief that feminism is causing the downfall of Western civilization.

In spite of these obvious differences in outlook, the men of the manosphere have a deep fascination with Stoic philosophy. Millennia-old Stoic texts are touted as "life-changing" and "mind-blowing" in glowing book reviews on Red Pill websites. Translations of the writings of Marcus Aurelius and Epictetus appear frequently in recommended reading lists for aspiring alpha males. There are threads requesting advice on how to apply Stoic principles to dating on message boards such as the r/theredpill subreddit. Even Valizadeh, since publishing the rant I quoted above, has embraced the philosophy to the point of writing on his blog, in a 2015 review of Seneca's *Letters to Lucilius*, that Stoicism "is the principal philosophy that I choose to live life by. It should be no surprise that this book is one of the most valuable in my library."[2]

Since Epictetus argued that other people's behavior and ideas are "not within our power," it is difficult to imagine a less Stoic pastime than ridiculing and attacking feminist writers for their ideas and physical appearances—an activity in which Valizadeh continues to take part. How can he and his followers embrace such behavior while considering themselves enthusiastic adopters of Stoicism?

This seemingly irreconcilable contradiction in fact reveals a key element of how ancient texts are used within the Red Pill community. Instead of deeply and productively studying Stoicism, its members use a simplified version of the philosophy to celebrate what they perceive as stereotypically masculine traits, including men's supposedly inherent superior capacity to use reason to control emotion. Via ancient Stoic ideas, these men appropriate the authority of the classical world to lend heft to their rhetoric and fuel their belief in their own superiority.

Stoicism is rising in popularity, and not just in far-right online communities. Recent years have seen many new publications about Stoicism, including William Irvine's *A Guide to the Good Life: The Ancient Art of Stoic Joy* and articles such as those Massimo Pigliucci has written for the *New York Times*.[3] There are robust online Stoic communities, such as a Stoicism subreddit with over forty-five thousand subscribers and the *New Stoa* website, which has more than two thousand registered users. Those interested in Stoicism can listen to podcasts, attend conferences, take part in an annual Stoic Week, or even watch Stoic stand-up comedy sets by Australian comedian Michael Connell.

What is causing the efflorescence of this ancient philosophy? As the classicist Chiara Sulprizio asked in 2015, in the very title of her article, "Why Is Stoicism Having a Cultural Moment?" Sulprizio suggests a number of Stoicism's elements make it particularly resonant in the modern era: a similarity with the self-help trend toward mindfulness and gratitude, compatibility with "New Atheism" and broadly defined spirituality, and the suitability of Stoicism's emphasis on cosmopolitanism to our hyper-connected world. The widespread embrace of Stoicism, as Sulprizio sees it, is an unquestionably positive and healthy phenomenon: "If this approach worked for past practitioners of Stoicism, why can't it work for us?"

But Irvine, Pigliucci, and Sulprizio—and others who celebrate the rise of Stoicism as a popular self-help philosophy—neglect to engage with the popularity of Stoicism in antifeminist internet communities.[4] That oversight is unsurprising, since these writers—and any others who have dedicated time to studying ancient Stoic texts in more than a superficial manner—know the Stoics believed that virtue has no gender and that men and women are both capable of using reason to determine what virtuous action would be. Stoicism, the philosophy beloved by the Red Pill, has a reputation among scholars as one of the most feminist of all ancient philosophies. The earliest Stoics even believed in radical gender equality. So mainstream writers advocating Stoicism may see its blossoming in antifeminist contexts as the result of a fundamental misunderstanding or as an anomaly and therefore disregard it.

But a closer reading of those same ancient texts reveals that the attitudes ancient Stoic writers held about gender and virtue were more complex and less progressive than they at first appear. These texts frequently offer seemingly egalitarian ideas but cloak them in language riddled with misogynistic undertones. So antifeminist readings of these texts in Red Pill communities should not be dismissed as misreadings or shallow interpretations: they may be responding to and drawing on parts of Stoicism that advocates of the philosophy would prefer to ignore. Deconstructing Red Pill uses of Stoicism is both an effective path into the community's broader ideology and an object lesson in the challenges of bringing the ancient world to bear on gender politics in the twenty-first century.

This chapter has two aims. The first is to analyze what is at stake for the men of the Red Pill when they publicly profess to study Stoicism. How does calling upon ancient philosophy work for them as a rhetorical strategy? The second aim is to situate the Red Pill interpretation of Stoicism in the broader context of a philosophy with complex and often contradictory approaches to gender, sex, and social justice. Through understanding how and why Stoicism appeals to the reactionary tendencies of the men who frequent Red Pill websites, we can also determine what in the philosophy can help promote progressive gender politics and social activism.

What Is Stoicism?

Scholars divide Stoic inquiry into three broadly defined and frequently overlapping categories: logic, physics, and ethics. Historians of philosophy have also often divided the major thinkers of ancient Stoicism into three chronological periods, sometimes called the Early Stoa, Middle Stoa, and Late Stoa. The first period begins with Stoicism's founding around 300 BCE in Athens and ends in the late second century BCE; from that point the Middle Stoa period, the era of the philosophers Panaetius and Posidonius, continues until the center of the philosophy (and that of the other major philosophical schools, including the Academics and the Epicureans) shifts to Rome in the first century BCE.

Red Pill writers who discuss Stoicism are interested only in Stoic ethics from the Late Stoa. That limited focus is entirely natural, considering that there are very few sources available to us from the earlier periods, and those few sources are fragmentary and only translated from the original Greek in texts aimed at an academic audience, such as A. Long and D. Sedley's two-volume *The Hellenistic Philosophers*.[5] It is therefore highly unusual to see references on Red Pill sites to the great Stoic thinkers of the Early Stoa, such as Zeno and Chrysippus, or the Middle Stoa, such as Panaetius and Posidonius.[6]

Most of the study of Stoicism in the Red Pill involves a few influential bloggers reading Stoic works from the Late Stoa in translation, then either distilling the primary concepts for their readers or offering a list of decontextualized quotations. These simplified Stoic ideas are then disseminated more broadly. A strong grounding in the history and theory of Stoicism will therefore not provide much insight into what Stoicism *means* to the Red Pill community. A sense of the development of the philosophy over the nearly five hundred years between its foundation and the death of Marcus Aurelius is not a true prerequisite for understanding how Red Pill writers appropriate the philosophy and cloak their ideology in terms borrowed from the ancient world. Nor will reading Marcus Aurelius in the original Greek shed light on how a list of quotations from a translation of his *Meditations* is used. For the first aim of this chapter—to determine what is at stake for far-right online communities in celebrating this ancient philosophy—we could safely ignore the vast majority of Stoic thought, including the entirety of Stoic logic and Stoic physics.

Nevertheless, a basic knowledge of the philosophical school's primary concepts, figures, and texts can reveal how Red Pill writers simplify and distort Stoic ideas. It will also help with the chapter's second aim of contextualizing Red Pill Stoicism within the larger scheme of Stoicism and contemporary gender politics. I will therefore provide a very abbreviated overview of Stoicism, with more material in the notes for interested readers.

Zeno, the founder of Stoicism, was born in Citium, a city in Cyprus. After moving to Athens, he is said to have studied under the Cynic philosopher Crates. The Cynics were primarily known for their extreme rejection of social norms: Diogenes, Crates's teacher, was a strict ascetic who supposedly lived for a time in a large *pithos*, or ceramic jar, in the marketplace and is said to have wandered during the day holding a lantern, claiming to be looking for a truly honest man. Cynicism had a profound influence on Stoicism, and Zeno seems to have agreed with the Cynics that social norms have no inherent moral virtue.

Despite the similarities between the two philosophies, Stoics were in general more likely to conform to prevailing standards for behavior and conduct.[7] Ancient sources relate anecdotes about Diogenes flaunting his disdain for social mores by urinating, defecating, and masturbating in public places. While Zeno likely would have agreed with his teacher's teacher that such activities were not morally wrong, he and his Stoic followers did not seem to feel a moral imperative to flout social norms so flagrantly.[8] It has likely been a factor in Stoicism's continuing popularity that the philosophy does not require its adherents to adopt grossly unconventional modes of behavior.

Zeno and his disciples became known as Stoics because they gathered to discuss their ideas in the *Stoa Poikile* (Painted Stoa), a structure on the north side of the Athenian *agora* (marketplace) built in the fifth century BCE. A stoa is a covered walkway or portico, often with a wall on one side and an open colonnade on the other. The paintings that gave the *Stoa Poikile* its name depicted historical and mythical battles, including an Amazonomachy—the battle between Theseus, the king of Athens, and the female warriors of legend.

Only fragments of Zeno's extensive writings have survived. After his death in 262 BCE, the meanings of those writings became a major subject of debate among the Stoics. The chief point of contention throughout the centuries was the precise meaning behind Zeno's statement that the main goal (*telos*) was to live consistently (*homologoumenōs*).[9] The difficulties with interpreting Zeno's ideas led to conflict between Cleanthes, Zeno's successor as scholarch (the head of the Stoic school), and Cleanthes's own successor Chrysippus. Of the two, Chrysippus was more important to the history of the school and also more prolific, producing more than seven hundred volumes, although Cleanthes composed the only surviving early Stoic fragment of any length, his *Hymn to Zeus*.

Since Stoic thought was passed down and debated by a multitude of opinionated scholarchs, the philosophy's major fields of inquiry—Stoic logic, Stoic physics, and Stoic ethics—are not monolithic entities. Each encompasses controversies debated passionately for centuries. Stoic logic (*logikē*) encompassed not just what we would think of as formal logic, but also rhetoric, grammar, and epistemology.[10] Stoic physics (*physikē*) is a kind of natural philosophy with metaphysical elements, including the study of the nature of matter, the soul, the divine, and the lifecycle of the universe.[11] Stoic ethics (*ethikē*) was aimed at determining what constituted virtuous action and, crucially, which acts were acceptable but not necessarily virtuous.

Stoic ethics in general is formulated around the idea that the only true good is virtue; anything not virtuous is not inherently good. According to Zeno, health and wealth are not inherently good, but rather merely preferable. Actions aimed toward obtaining these are therefore appropriate *only* as long as they do not conflict with virtue. Anything that does not either tend toward or obstruct virtue is indifferent (*adiaphoron*).

A perfectly wise person—a sage—will take perfectly right actions (*katorthōmata*). However, even one who has not attained sagehood can devote himself to taking only appropriate actions (*kathēkonta*). The goal was a lack of passion (*apatheia*). Passions (*pathē*) are the result of giving assent to or endorsing false impressions—the result of misunderstanding things that are preferable, or indifferent, as things that are inherently good.[12] For example, something that negatively affects one's prosperity, while not preferable, is neither inherently bad nor worth upsetting one's *pathē*. Almost every Stoic text that can still be read today is a practical ethical work dedicated to advice on how not to give assent to these false impressions. Marcus Aurelius writes, "Today I escaped from all disturbances—or rather I cast them all out, because they were not outside of me, but in my own perceptions" (*Meditations* 9.13).

Between Zeno and Marcus Aurelius, however, lies a span of nearly five hundred years, and the focus of Stoic philosophers shifted during that time. After Sulla's capture and sacking of Athens as part of the Mithridatic War in 89–84 BCE, the city began to lose its status as the home of all philosophical schools. What followed was a period of decentralization and philosophical diaspora. As a result of that major shift, Imperial Roman Stoicism is markedly different from earlier iterations of the philosophy. In this period, Stoic logic and physics, which were both major concerns in earlier periods, are substantially ignored in favor of Stoic ethics.[13] Marcus Aurelius—who is admittedly not a pure example of late Stoicism, since he was heavily influenced by other philosophies—asserts that he is thankful he did not fall into an obsession with logical syllogisms or physical phenomena such as meteorology (*Meditations* 1.17).[14]

Almost all of the widely available texts about Stoicism were written by only a few authors: the aforementioned Marcus Aurelius and Seneca along with Cicero, Musonius Rufus, and Epictetus. By virtue of being the only Stoic thinkers whose texts have survived (at least, in any substantial form) into the present day, these men have cast a long shadow over Stoicism as it is understood by everyone aside from philosophy scholars.

The goal of Stoic ethics was to become a sage (*sapiens* in Latin), a person with a perfect understanding of virtue who only takes right actions. The ancient Stoics, along with modern scholars of Stoicism, do not universally agree on whether it is possible to achieve true Stoic sagehood: the best most people can hope for is to continue to progress along the path to sagehood and become a *proficiens*, one who has started (*proficere*) the journey and is making progress but has not yet reached its conclusion.[15] The Stoic writers whose texts are widely read today, including those discussed on Red Pill websites, were *proficientes*, not sages.

These men were proficient in Stoicism in the modern sense, but they were also self-consciously flawed, and the authorial personas they project in their writings on Stoicism have had a significant impact on how the philosophy is understood today. Many of the texts they wrote actually perform the journey toward sagehood by showing the reader isolated incidents of actively working through and trying to overcome false impressions. The quirks and interests of these writers have had a long afterlife in the reception of Stoicism, especially within the manosphere, where the study of individual exemplary historical figures is typically favored over broader historical trends in the ancient world.

Marcus Tullius Cicero (106–43 BCE), the great Roman politician, orator, and polymath, is the earliest extant thinker from whom we have surviving, complete treatises about Stoicism. Cicero was a prolific writer who left behind a vast body of work, including speeches, essays, some poetry, and extensive personal correspondence. A few of these texts focus on Stoicism, including his *Tusculan Disputations* (*Tusculanae Disputationes*), *Stoic Paradoxes* (*Paradoxa Stoicorum*), and *On Duties* (*De Officiis*). He is an invaluable source for his summaries of lost Greek Stoic texts.

Cicero was not, however, exclusively or even primarily a Stoic. He claimed to be primarily an Academic with some Stoic leanings, and he exemplified philosophical eclecticism, not orthodoxy. Cicero occasionally writes critically of Stoicism, as in his fictional debate with Marcus Porcius Cato in the third and fourth books of his *De Finibus Bonorum et Malorum* (*On the Ends of Good and Evil*).[16] This ambivalence provided the foundation for critiques of Stoicism that are still influential today.

Regardless, Cicero offers a valuable model for how even imperfect Stoics can benefit from studying the philosophy. His most notable (yet arguably productive) failure at assimilating Stoic thought was his tremendous grief after the death of his daughter Tullia. Stoics preach that the death of a child is not evil, but indifferent: Epictetus praised the truism, variously attributed to the Athenian lawmaker Solon and the pre-Socratic philosopher Anaxagoras, that when a parent suffers the death of a child, the correct response is to say, "I knew I had fathered a mortal" (Epictetus, *Discourses* 3.24.105). Cicero's *Tusculan Disputations* is an extensive, philosophical, and only nominally successful attempt to come to terms with that indifferent (in Stoic terms) event.

The next of the Stoic writers whose work survives was Lucius Annaeus Seneca, also called Seneca the Younger. Son of the noted rhetorician Seneca the Elder, Seneca the Younger was born sometime between 8 and 1 BCE. He had the dubious honor of running afoul of three emperors: Caligula, who nearly executed him; Claudius, who exiled him to Corsica for adultery with the emperor's niece; and Nero, who eventually ordered him to commit suicide. Seneca was an adviser to Nero from 54–62 CE; from 62–65 CE he was in retirement, and this period seems to be when he produced many of his philosophical writings about the importance of emotional control, including his *Letters to Lucilius*. In 65 CE, he was implicated in the Pisonian conspiracy to assassinate Nero, at which point Nero ordered him to kill himself. The later historian Tacitus assimilates this noble death scene to the forced suicides of other philosophers before Seneca, including that of Socrates.[17]

Seneca's works—which include essays, tragedies, and *Letters to Lucilius*, a set of philosophical letters to a younger friend—are some of the most enjoyable Stoic texts to read, because Seneca does not share the dour outlook of later writers, such as Marcus Aurelius. Seneca was fabulously wealthy and powerful: Cassius Dio criticized him for supposedly bankrupting Britain by calling in his debts, and claimed that Seneca's net worth was over 300 million sesterces (61.10.3). Although it is extremely difficult to translate sesterces to modern currency, most estimates would put that amount at over a half a billion dollars.

Seneca repeatedly justifies his expensive tastes, saying that as long as he recognizes those pleasures are not good in themselves—as long as he *could* give them up with equanimity—he is not acting in a way that would be inconsistent with Stoic thinking. He imagines and anticipates these criticisms in *De Vita Beata*:

> Why do you believe that you need money, and why are you disturbed when you lose it? Why do you shed tears at the death of your wife or your friend? Why do you care about the malicious stories people spread about you? Why is your country estate outfitted so much more elaborately than it needs? Why do you not dine according to your precepts? Why is your furniture so expensive? Why do you drink wine that is older than you are? Why do you have elaborate landscaping? Why do you plant trees that bear no fruit other than shade? Why does your wife wear earrings that cost as much as an expensive home. . . . For now, I will answer this: "I am not a sage [*sapiens*], and I will not become one just to satisfy your spite. Do not demand that I be equal to the best men, but only that I be better than the worst. It is enough for me if each day I remove some of my vices and correct some of my faults. I have not arrived at perfect understanding, and I never will . . . but compared with your slow feet, I am a racer." (17)

The obvious value that Seneca placed on wealth—an indifferent, in Stoic terms—has led to frequent accusations of hypocrisy.[18] These are unsurprising, yet also somewhat unfair, because Seneca admits freely that he is not a sage and there is not much point in criticizing him for imperfections he never denies having.

Seneca's contemporary Musonius Rufus, another major Stoic thinker, also came from an equestrian family—the lower of the two aristocratic classes, below the senatorial class—practiced Stoicism, and suffered under the reign of Nero. He fared somewhat better than Seneca did: Musonius was merely exiled, not executed or forced to commit suicide. And Musonius did not believe that exile was a terrible fate, unlike his fellow Stoic writers: Cicero complained bitterly in his correspondence during his exile, and Seneca's letter of consolation to Polybius is widely believed to be a thinly disguised attempt to beg Claudius to allow him to return.[19] In contrast, Musonius dedicated his ninth lecture to the subject of why exile is not an evil. Musonius's lectures come to us secondhand, transcribed by a student and then collected by Stobaeus, a late-antique Macedonian anthologist whose life we know practically nothing about. Like his pupil Epictetus, Musonius followed in the footsteps of Socrates in not producing written works of philosophy.

Musonius is primarily known for advocating teaching philosophy to women as well as men. His third and fourth lectures are devoted to this subject. Stobaeus quotes him making this argument:

> For one thing, he said, women have received from the gods the same reasoning power as men—the power which we employ with each other and according to which we consider whether each action is good or bad, and honorable or shameful. . . . In addition, a desire for virtue and an affinity for it belong by nature not only to men but also to women: no less than men are they disposed by nature to be pleased by noble and just deeds and to censure things opposite these. Since this is so, why would it be appropriate for men but not women to seek to live honorably and consider how to do so, which is what studying philosophy is? (3.1–2)[20]

While this assertion that "an affinity [for virtue] belong[s] by nature not only to men but also to women" may seem radically feminist for his time, Musonius's view of appropriate female behavior was actually fairly conventional, as I will discuss later. His argument is instead that philosophy can help women to perform their socially ordained roles as wives and mothers. Musonius's version of Zeno's consistency includes an understanding that virtue may look very different for people of different genders and classes.

Epictetus was Musonius's student, and their ideas are strikingly similar. Little is known about Epictetus's life. He is thought to have been born around 50 CE in Hierapolis, modern-day western Turkey. A slave until the age of thirty, he began attending Musonius's lectures while still enslaved (*Discourses* 1.9.29). Celsus attributes his lameness, referred to in his *Discourses*, to having had his legs broken by his master, often believed to be the freedman Epaphroditus (1.16.20). However, there is no clear reference to either his former master or to the source of his lameness in his works, which were written down by his student Arrian, a historian and philosopher.[21]

The last of the remaining major Stoic writers, Marcus Aurelius, is also perhaps the most influential, especially throughout the Red Pill community.[22] His appeal undoubtedly owes a great deal to his being not only famously learned and wise—as the manosphere blog *Illimitable Men* puts it, "introspective and inquisitive"—but also perhaps the most powerful man in the world.[23] Marcus Aurelius ruled the Roman Empire from 161–180 CE, leading it through conflicts with the Parthian empire in the East and the Germanic tribes in central Europe, as well as a significant revolt led by his own general Avidius Cassius.

In the latter part of his reign, Marcus Aurelius kept a journal, not of events, names, and places, but of philosophical ideas. Although this journal is usually called *Meditations*, its original Greek title seems to have been *To Himself* (Τὰ εἰς ἑαυτόν), and all signs indicate it was indeed written for his own eyes and no others. With the exception of the first book—a list of those to whom he owes gratitude, with explanations for the benefits they have given him—the rest of the work is disorganized and obscure to modern readers. He refers to individuals and events that are now unidentifiable, and he makes no attempt to package his often-depressing thoughts in clever maxims or entertaining anecdotes to appeal to readers, as Seneca did.[24] One entry on the human condition reads, "The smell of a goat, viscera in a sack. Look at it clearly" (8.38). Nevertheless, he is excruciatingly, unswervingly honest with himself about his own failings in a way that has proven compelling for centuries of readers. And his reminders about how to endure the failings of others with equanimity seem as though they would resonate with any reader in any era: "In the morning, tell yourself this: I will encounter people who are meddlesome, ungrateful, insolent, dishonest, malicious, and bad-tempered. Ignorance of good and evil makes them this way" (2.1).

The New Stoa Poikile

From the abbreviated summary above, it may not be obvious why Stoicism has proven so attractive to far-right, antifeminist online communities. But while Red Pill writers must gloss over quite a few ideological differences—and ignore centuries of Stoic debate on fine points of logic and ethical theory—to market themselves as the inheritors of the philosophy, they are willing to rationalize those discrepancies to access what they perceive as Stoicism's main benefit: its utility as a self-help tool.

Many men are attracted to the Red Pill not only because it provides them with a community of other like-minded men, but also because they are seeking advice on how to improve themselves. The advice dispensed on these fora tends to fall into a few distinct categories: physical fitness, such as weightlifting and martial arts; professional concerns, with a special focus on navigating politically correct office environments and becoming economically self-sufficient; and dating advice, or "how to have game," a topic I will return to in the next chapter.

Stoicism focuses explicitly on self-improvement, so it can be blended easily into the self-help aspect of Red Pill communities. For example, Seneca asks in *De Ira* (*Concerning Anger*): "Which negative trait of yours have you eradicated today? Which vice did you stand up to? In what way are you better?" (3.36). In addition to encouraging Stoic *proficientes* to strive to improve themselves, the Stoics also had concrete tools they used to increase their equanimity and guard against false impressions. The most famous of these is the *praemeditatio malorum*, a practice of envisioning possible outcomes so as to be inoculated against negative reactions.[25]

The Red Pill emphasizes Stoicism's practicality in nearly every article about the philosophy, list of quotations from Stoic thinkers, and appearance of Stoic texts on recommended reading lists for aspiring alpha males. *Illimitable Men*, a blog that uses a more literary and philosophical approach than most manosphere sites, lists Marcus Aurelius's *Meditations* second in the top ten "books for men," describing it as "helpful as a spiritual guide to dealing with, and perceiving life."[26] *Meditations* also appears on a list of "Comprehensive Red Pill Books" on the Red Pill subreddit, where it is described as "a very simple pathway to practical philosophy."[27] In a review of Epictetus's *Enchiridion* on *Return of Kings*, Valizadeh praises Stoicism and claims that "Stoicism will give you more practical tools on approaching life and dealing with its inevitable problems."[28] Valizadeh also wrote a 2016 review of *Meditations* with the effusive title "Marcus Aurelius' *Meditations* Is the Best Manual We Have on How to Live."[29] An article in the online alt-right journal *Radix*, "Viewing Stoicism from the Right," admits that "Stoicism gives you practical directions to follow."[30] On the blog *Black Label Logic*, which takes a rhetorical and analytical approach to Red Pill ideology, an article about Seneca and Machiavelli instructs readers that "Seneca and Marcus Aurelius give you the tools to control how the world affects you."[31] Again and again, these writers praise the *utility* of Stoicism in the present day, de-emphasizing the vast majority of less practical elements in Stoic thought.

This tendency among Red Pill Stoics is part of a bigger trend in twenty-first century popular Stoicism more generally, the rebranding of the philosophy as a "life hack" (or "mind hack"), a strategy or technique that increases efficiency and productivity. Some major figures in the Stoic community, including Massimo Pigliucci, warn against this use of Stoicism: Pigliucci cautions that "adopting a philosophy of life—or a religion, which is a type of life philosophy—is a bigger deal, and cannot be reduced to life hacking."[32] Nevertheless, this use of Stoicism is on the rise, especially in Silicon Valley.[33]

One of the foremost proponents of Stoicism as a life hack is Ryan Holiday, a former marketer and author of the 2012 book *Trust Me, I'm Lying: Confessions of a Media Manipulator.* Holiday has since shifted the focus of his career—a profile in the *New York Times* called him "a charismatic public-relations strategist turned self-help sage"—publishing two books meant to be modern advice guides based on Stoic principles: *The Obstacle Is the Way: The Timeless Art of Turning Trials into Triumph* (2014) and *Ego Is the Enemy* (2016). Holiday, an energetic proponent of Stoicism as a life hack, has written extensively on the topic both in his books and for online publications, including his personal blog, *The Guardian, Business Insider, Thought Catalogue,* and countless others. He claims to have discovered Stoicism when Dr. Drew Pinsky, the celebrity doctor, recommended the works of Epictetus to him. After Holiday's associate Tucker Max, the famous bro-turned-relationship-guru and author of *I Hope They Serve Beer in Hell* (2006), also suggested that Holiday read *Meditations,* Holiday decided to do so; he claims, "My life has not been the same since."[34]

Although the announcement that Holiday would be a keynote speaker at STOICON 2016 met with some suspicious murmurs on Stoic blogs and fora, in some ways he is an ideal mouthpiece to bring Stoicism to unexpected and untraditional audiences. Many of those who are interested in the philosophy, including the men of the Red Pill, need someone to repackage it for them in a more easily digestible form. Although that work can certainly be done by philosophers and scholars, Holiday benefits from being untouched by the perceived ivory-tower impracticality of academia—a distance that is especially necessary for reaching the men in the Red Pill, who generally believe that university professors are out-of-touch leftists. As a self-taught Stoic who left college without graduating, Holiday has unique appeal.

While Holiday is not quite a member of the Red Pill community, he is certainly skilled at creating and promoting the kind of content that appeals to the men who frequent its websites.[35] He has worked in marketing for many authors beloved by the Red Pill, including not only Tucker Max but also Robert Greene, author of the book *The 48 Laws of Power*—the only volume ranked ahead of *Meditations* on *Illimitable Men*'s list of "books for men."[36] Greene's works are a fixture on pickup artist reading lists, and James "Roissy" Weidmann quotes them frequently on his site *Chateau Heartiste*. Holiday also approvingly quotes Nassim Nicholas Taleb, whose Stoic-influenced book *Antifragile* is enthusiastically praised by many in the Red Pill, and likewise, Holiday's own books have been reviewed by many Red Pill sites. Despite not associating himself directly with any Red Pill virtual fora, Holiday has done a masterful job, as befits a self-professed media manipulator, of ensuring that the men who subscribe to those sites will form a loyal part of his audience. The Red Pill prefers his work on Stoicism over that of reputable Stoic scholars such as Pigliucci.

The Obstacle Is the Way is not a book about ancient Stoicism; rather, it is itself a Stoic text, following the model of its ancient predecessors, and aims to teach its readers how to conquer obstacles. Holiday structures the book into three sections—Perception, Action, Will—from a similar tripartite division in Marcus Aurelius's thinking (*Meditations* 7.54). But even though Stoicism and Marcus Aurelius provide a structure and frame for the book, Holiday does not expend much energy analyzing Stoic texts, preferring instead to make self-help suggestions buttressed by quotations from Stoic thinkers and anecdotes from the lives of great men. Stoicism's influence on Holiday's *Ego Is the Enemy* is even subtler, although the book is still peppered with Stoic quotations.

It is telling that Holiday chooses to pre-digest and regurgitate Stoicism for his readers. Treating Stoicism as a quintessentially practical philosophy requires one to either divorce passages from their original context, so as to make them seem universally applicable (as Holiday does in his listicles on various websites), or to distance oneself from the text entirely, by illustrating Stoic principles through anecdotes about exemplary figures (as Holiday does in his books). These tactics cleverly mask the fact that, while much of what Marcus Aurelius and other Stoic thinkers wrote seems directly applicable to the practicalities of life in the twenty-first century, much of it seems equally foreign to our sensibilities and quite technical as well, especially Marcus Aurelius's frequent ruminations on the nature of the *logos* and its place in our lives. The same problem will arise in Chapter 3, when I analyze how pickup artists read Ovid's seduction advice. In both cases, there is a remarkable tendency to overstate how seamlessly the wisdom of ancient texts can be applied to today's world. Lessons from ancient texts *can* be applied to the present world, but rarely with ease, and ancient Stoic texts may require adaptation before they can become sources of practical advice for how to navigate the twenty-first-century world.

These distortions are particularly necessary to making Stoicism palatable for the Red Pill community, most of whose members are vocal nationalists who have completely bought into the idea of making America great again, because Stoicism is a cosmopolitan philosophy. Epictetus advised, following Socrates, "When someone asks you what country you are from, never say that you are Athenian or Corinthian, but rather that you are a citizen of the world [*kosmios*]" (Epictetus, *Discourses* 1.9). Epictetus's "citizen of the world" concept is shared by all ancient Stoics. Musonius's argument that exile is an evil depends on the idea that "the world is the common fatherland of all human beings," and Marcus Aurelius writes, "As Antoninus, my city and fatherland are Rome. But as a human, they are the world. So what is good for me must also be good for both of these" (Musonius, *Lectures* 9.2; Aurelius, *Meditations* 6.44). Many of the factions within the Red Pill, particularly the white nationalists of the Alt-Right, would call any attention to the interests of other nations a sign of being a cuck.[37] Multiculturalism is the avowed enemy.

Marcus Aurelius's responsibility to weigh his decisions against their effects on the community also poses a bigger problem for Holiday's use of Stoicism as a life hack: it calls into question the elevation of self-improvement above the pursuit of meaningful systemic change. Holiday's version of Stoicism encourages readers to use Stoic maxims for the un-Stoic goal of improving their station in life, without necessarily considering the good of the community. In the introduction to *The Obstacle Is the Way*, while explaining why it can be helpful to study how successful men have dealt with their problems, Holiday writes, "Whether we're having trouble getting a job, fighting against discrimination, running low on funds, stuck in a bad relationship, locking horns with some aggressive opponent, have an employee or student we just can't seem to reach, or are in the middle of a creative block, we need to know that there is a way. When we meet with adversity, we can turn it to advantage, based on their example."[38] "Stuck in a bad relationship" seems on par with "fighting against discrimination" to Holiday, who claims to have discovered Stoicism while fighting depression after a bad breakup. Similarly, many pickup artists and Men Going Their Own Way would argue that the dating world as it exists today is rigged against men, so the process of finding a girlfriend *is* a kind of "fighting against discrimination." But that is only true for heterosexual, white, cisgender men who never face the kind of discrimination that does not allow one to turn obstacles into opportunities.

Later in the book, Holiday is even more explicit about how little attention he believes people should pay to discrimination: "The point is that *most people* start from disadvantage (often with no idea they are doing so) and do just fine. It's not unfair, it's universal. Those who survive it, survive because they took things day by day—that's the real secret."[39] Holiday puts racism, sexism, ableism, homophobia, and a host of other prejudices into a box, labels it "disadvantage," and then makes it vanish by proclaiming disadvantage universal to the human condition. Holiday and Marcus Aurelius both write from positions of privilege, but Marcus Aurelius repeatedly asserts the importance of remembering that one is responsible to one's fellow men, an attitude hardly in evidence in Holiday's work.

The sexual ethics of Stoicism are also incompatible with those of the manosphere. Different factions within the manosphere have different approaches to sex, ranging from the pickup artist community's single-minded focus on obtaining it to the Men Going Their Own Way and their "marriage strikes." Stoicism fits poorly with both of these outlooks, even though one small blog, *Rex Patriarch*, asks, "Is MGTOW the Idea of Ancient Stoicism Repeating Itself?"[40] Zeno, following his Cynic roots, thought the ideal city would outlaw marriage entirely to avoid jealousy, but later Stoics from Antipater onward think that marriage is almost a prerequisite for philosophy.[41] Musonius argues in his twelfth lecture that men and women alike should only have sex within marriage; this belief—that adultery was a fault for both genders—was anomalous in a society that harshly punished female adultery but shrugged off male adultery.[42]

Furthermore, Marcus Aurelius views all pleasures with suspicion. He encourages himself, when looking upon sensual delights, to remember what he is really seeing: "The corpse of a fish, or of a bird or a pig. An excellent vintage of wine is grape juice, and rich purple clothes are sheep's wool colored with the blood of a shellfish. Coitus is just friction on your penis, a spasm, and the expulsion of some mucus" (*Meditations* 6.13). Marcus Aurelius also wrote that men should pray "not 'How can I sleep with her?' but 'How can I stop wanting to sleep with her?'" and thanks the gods that he did not lose his virginity too early (*Meditations* 9.40, 1.17). This sentiment is difficult to assimilate with threads on the Red Pill subreddit such as "How to Become Outcome Independent Using a Stoic Trick," an article with advice for how to maintain the upper hand sexually in a long-term relationship by training yourself not to care what your significant other thinks.[43] Another thread advises, "Men, you can never underestimate the power that stoicism and strength has on a woman. It can bring her from wanting to ruin you to busting her ass to get validation from you like a child would."

Red Pill writers who endorse Stoicism have to go to considerable lengths to elide these less congenial aspects of the philosophy. Yet they expend the effort to overemphasize Stoic texts' usefulness and practicality in order to make them more palatable.

The real answer to the question "Why Stoicism?" can be found not in the writings of Holiday, but in those of Red Pill community members who devote less energy to Stoicism's practical aspects and tout its intellectual benefits instead. These authors are less likely to share lists of quotes and more likely to compose essays about the value of the dedicated study of Stoic texts. In their musings about Stoic philosophy's value to the modern man, they reveal why Stoicism continues to exert such appeal over the men who have "swallowed the red pill": in short, it justifies their belief in the intellectual superiority of white men.

One of these writers is among the most educated and articulate contributors to the site *Return of Kings*: Quintus Curtius, who takes his name from the ancient historian Quintus Curtius Rufus, who wrote a history of Alexander the Great in the first or second century CE. Quintus, who remains completely pseudonymous—unusual for Red Pill "thought leaders" trying to build personal brands but common on Red Pill sites more generally—has eclectic tastes in history, philosophy, and literature. For a few years he wrote a weekly column on *Return of Kings*, and he runs his own blog, *Fortress of the Mind*, about European history, European philosophy, and "Great Men."

Although Quintus Curtius does not explicitly use Stoicism as a self-help philosophy, he does frame his larger project as one promoting self-improvement through education:

> Readers will note that I have long emphasized historical and philosophical topics as ways to make larger points. There is a deliberate reason for this. By invoking the past, I have tried to remind readers of the glories of leadership, character, and masculine virtue that can change their lives. By bringing up the past, a time before masculine virtues were shamed and punished, we remind readers of the glories that will be theirs if they follow the right paths.[44]

This approach is highly individualistic, as is "neomasculinity," the philosophy Quintus espouses on *Return of Kings*. He encourages his readers to focus on the self as the locus of change rather than critiquing larger structures of oppression and discrimination. He hints at the existence of those structures, especially the harmful gender norms that contribute to an atmosphere in which "masculine virtues [are] shamed and punished," but urges readers to ignore social change in favor of a blinkered fixation on self-improvement.

Quintus claims to be a former Marine and a lawyer, a crucial (if unverifiable) detail because, as he argues on his personal blog in a promotional post for his translation of Cicero's *De Officiis*, his purported trial experience gives him privileged access into the mind and language of Cicero, who wrote many of his most famous speeches for the law courts.

> We must remember that Cicero was always the lawyer, arguing his points with force, conviction, and clarity. . . . He knows that juries, like book readers, need emotional connection, summary, and repetition. Some translators, because they are not trial attorneys by profession, entirely miss this point. But those of us who have tried cases before juries—and I am not aware of any other translator of *On Duties* who was also a trial lawyer—see the method and purpose in Cicero's rhetoric.[45]

Like Holiday, Quintus constructs his authority through opposition to traditional scholarship. His own experience as a legal professional makes him better suited to understanding the minds of ancient lawyers than are classicists who study Cicero.

However, unlike Holiday—who told a crowd at STOICON 2016 that Stoicism was "a philosophy designed for the masses"—Quintus praises Stoicism for its rarified appeal. He writes,

> Some of Cicero's philosophical works have made me appreciate that Stoicism, *while not a philosophy for the masses*, attracted the best men everywhere it took root; and it was an honest attempt to impart a moral code to the ruling classes of the ancient world before the advent of Christianity. It was the most far-reaching and influential of the pre-Christian philosophical systems in the West, and it still attracts men today for the masculine resonance and austere grandeur of its precepts.[46]

Quintus's endorsement of Stoicism betrays his elitism ("the best men") and sexism ("the masculine resonance . . . of its precepts").

His claim that Stoicism "still attracts men" implies that Stoicism is less well-suited to female audiences, while remaining enticingly generalized: although Stoicism was previously the province of the "ruling classes," it is now attractive to "men," a demographic to which he attaches no qualifying adjective. By studying Stoicism, all men today can assimilate themselves to the best ancient Romans.

While Quintus and Holiday use opposing rhetoric to frame the appeal of Stoicism to their audiences, close reading reveals that Holiday's thinking is infected by the same sexism and elitism that Quintus displays, albeit less explicitly. As I argued above, Holiday's entire approach of teaching Stoicism through the examples of remarkable historical figures is subtly elitist, prizing success without taking into account the structures of privilege and oppression that make success more easily accessible to some than to others. Unsurprisingly, many of his exemplars of Stoic ideals are white men born to wealth. He mentions only two women, Laura Ingalls Wilder and Amelia Earhart, although he claims to focus on "great men and women of history."[47] Perhaps this is why Sandy Grant, a philosophy scholar at the University of Cambridge, has deemed Holiday's work "bad pop psychology of a comically macho bent for sale to entitled and arrogant successniks."[48]

The combination of practicality and elitism that Holiday and Curtius display has proven irresistible for the men of the Red Pill. Ancient Stoic texts reinforce their belief that their ideology—namely, that male self-actualization and self-improvement is the highest possible good—has deep roots and undeniable intellectual credibility.

Stoic Feminism and Stoic Sexism

One idea Stoicism reinforces for the Red Pill community is that men are superior to women because they are by nature more rational and less emotional. In some ways, this use of Stoicism is a distortion of the ancient Stoics' stated views about gender and rationality: because the Stoics believed in the theoretical equality of the genders, it is surprising to see Stoicism used to promote retrogressive gender norms. However, hidden just beneath the surface of Stoicism's apparent protofeminism is a gender politics easily adaptable to Red Pill ideology.

The Stoics believed that, in theory, both men and women were capable of rationality and of becoming sages. The Roman Stoic writer Seneca wrote as much in a letter consoling Marcia, the daughter of a prominent historian, after the death of her son Metilius:

> I know what you would say: "You have forgotten that you are consoling a woman, and the examples you are using are those of men." But who said that nature has been stingy with the characters of women, or that nature has withdrawn their virtues? Believe me, if women wish to, they have vigor and skill at acting justly equal to men, and if they have experience then they endure pain and labor as well as we do. (*Consolatio ad Marciam* 16.1)

On the other hand, it is difficult to deny that misogyny is deeply ingrained in the thinking of many of our extant Stoic texts—the same misogyny that, as I argued in the previous chapter, makes the men of the Red Pill feel especially at home reading classical texts. Even Seneca's address to Marcia suggests that denigrating women was commonplace. As Lisa Hill writes,

> Any contemporary reader of Stoic prose will be struck by the cavalier sexism which pervades their work. Though stylistic misogyny is the norm rather than the exception in classical writing, one somehow has higher expectations of thinkers avowedly intent on ignoring social distinctions and otherwise famed for their tolerance and humanity.[49]

Nevertheless, on some counts, the reputation Stoicism enjoys as an unusually feminist philosophy is deserved. An egalitarian attitude toward the sexes is a component of Stoicism that goes back to the philosophy's beginning, and evidence suggests that the Stoic belief in the possibility of female sages began in the early days of the school.

Cleanthes, Zeno's successor, wrote a work called "On the Position that a Man's and Woman's Virtue Is the Same" (D. L. 7.175), but of the Stoic writers whose works remain, Musonius Rufus is most likely to be called a feminist.[50] Two of his lectures in particular support this thesis: his lecture arguing that women can also learn philosophy (lecture 3; ὅτι καὶ γυναιξὶ φιλοσοφητέον) and his lecture arguing that both sons and daughters should receive an education (lecture 4; εἰ παραπλησίως παιδευτέον τὰς θυγατέρας τοῖς υἱοῖς). He argues in lecture 3 that women have the same *logos* as men, the same desire for virtue, and the same need to learn how to live appropriately. In lecture 4, he repeats some of the arguments from lecture 3 about women needing virtue and points out that gender is not a consideration in the training of animals, such as horses.

Although Musonius's ideas about female education may have seemed radical in his own time—and form a definite contrast with the Red Pill attitude toward female education, which is perhaps best summed up by a 2013 article by alt-right blogger Matt Forney, "The Case against Female Education"—Musonius's agenda in teaching women philosophy is that it would better prepare them to run a household and raise children.[51] According to Musonius, "both men and women must live in accordance with justice. A man would not be a good citizen if he is unjust, and a woman would not manage her household well, if she does not do it with justice" (*Lectures* 4.2).[52] He also assures his audience that a female philosopher would not wander around town like a male philosopher, but remain modestly at home and even breastfeed her own children (*Lectures* 3.6–7). And, as feminist philosopher Martha Nussbaum has pointed out, all of these texts discussing the possibility of female virtue are fundamentally antifeminist by virtue of being aimed at male audiences.[53]

Nussbaum concludes that considering Musonius's contingent circumstances—that he was Stoic and Roman—he was "in advance of Roman customs of the day" but that "there are aspects of Musonius's feminism that should be questioned by anyone with an interest in women's complete equality."[54] Even though some Stoics believed both men and women should receive an education and are both capable of virtue, they did not necessarily believe men and women should hold the same roles in society. Although some scholars, including Elizabeth Asmis, argue that Stoic marriage was a radically egalitarian construct, other Stoic texts suggest that the model was one of complementarity, not equality.[55] To be fair, this stance toward marriage is preferable to the views of Hesiod and Semonides mentioned in the first chapter of this book—namely, that wives are always a burden—but in the end, even Musonius's ideal marriage appears similar to Ischomachus's marriage in Xenophon's *Oeconomicus*. Most ancient male writers seem to have agreed that a woman's highest calling is to be a superlative homemaker.

Musonius also displays the basic assumption that has made Stoicism so attractive to the manosphere: that is, that men are more naturally suited to emotional control than women are. He reveals this outlook when discussing why it is inappropriate for men to sleep with their female slaves, a practice that was generally considered socially acceptable in the ancient world:

> In response to this, my reasoning is simple: if someone thinks that it is neither shameful nor unnatural for a master to consort with his own female slave (and especially if she happens to be unmarried), what would he think if his wife would consort with a male slave? Would he not think that this was intolerable, not just if a woman who had a lawful husband would submit to a male slave, but even if an un-

married woman would do this? And yet, no one will suggest that men should have a lower standard of conduct than women or be less able to discipline their own desires—that those who are supposedly stronger in wisdom would be bested by those who are weaker, or that those who are rulers would be bested by those who are ruled! Men should have a much higher standard of behavior if they expect women to follow them. (*Lectures* 12.4)[56]

Musonius's thinking here and elsewhere is consistent with the kind of assertions made by the manosphere in articles such as "6 Ways 'Misogynists' Do a Better Job at Helping Women than Feminists."[57] In addition to being predicated on the idea that men have superior emotional control and should serve as leaders—partly example, partly guardian—for women, such articles claim that the men on Red Pill sites care more about women than feminists do because they want women to be safe, healthy, and happy. They also claim the right to determine for women what safe, healthy, and happy mean.

The feminism of ancient Stoicism seems to extend only as far as acceptance that women can be rational because they too have a share of the *logos* and the belief that women should not be disrespected or abused. For example, Epictetus criticizes men for making women shallow and obsessed with their appearances: "From fourteen years of age onward, women are called the possessions of men. When they see that there is no other future for them than to have sex with men, they begin to put all of their focus into improving their appearances. It is worth our while to help them learn that they will be honored only for being well-behaved and modest" (*Enchiridion* 40). Epictetus blames men for female vanity and superficiality, and urges them to act in such a way as to encourage women to value virtue.

While Epictetus's prescription here is undoubtedly an improvement from the mindset he advocates against, it is not feminism in

any meaningful sense. It does not overturn many of the very restrictive ancient norms about appropriate female behavior—norms
the Red Pill praises. So while many men in the manosphere, especially the pickup artists, act precisely counter to how Epictetus recommends by treating women as sex objects, both would agree that
female behavior is, and should be, primarily reactive to and guided
by male behavior.

But the failures of Stoic feminism in the ancient world go beyond
believing that women belong in the home: Stoic texts are also filled
with insinuations that women are generally inferior to men. Cicero's ideal fellowship is one of "good men" (*boni viri*). Rather than
simply *boni* alone, which could refer to a collective group of good
men and women, he uses the emphatic *viri*, thus gendering the fellowship and revealing his preference for communing philosophically with his own gender (*De Officiis* I.55).[58] Seneca believes that
women have a low capacity for self-control (*Ad Helviam* 14.2). He
also praises Marcia for not showing the customary "weakness of
the feminine mind" (*infirmitate muliebris animi*), a claim that
finds echoes in the Red Pill community's favorite blue pill maxim,
"not all women are like that" (*Ad Marciam* 6.1). Epictetus, who
agrees with Musonius (quoted above) that women have difficulty
keeping their emotions in check, judges Epicureanism "not even appropriate for women" (οὐδὲ γυναιξὶ πρέποντα)—a clear indication
that gender *is* a consideration for him in the practice of philosophy
(*Discourses* 3.24.53; 3.7.20).

Furthermore, although female virtue is granted as a possibility,
virtue itself is always coded in ancient texts as male. The words for
virtue in both languages—*andreia* in Greek, *virtus* in Latin—
literally mean manliness.[59] Cicero and Seneca both use adjectives
meaning "like a woman," such as *muliebris* and *effeminatus*, to denote negative qualities, whereas adjectives meaning "like a man,"
such as *virilis*, have positive connotations.[60] Likewise, in *Meditations*, Marcus Aurelius insults an unidentified man's character as

"womanly" (4.28; θῆλυ ἦθος). Stoic texts in Latin exhort men not to exhibit *mollitia* (softness), a quality usually associated with women.[61] While the idea that male characteristics are superior to female ones is not explicit in Stoic thinking, it is deeply encoded in it.

On the one hand, Cicero and Seneca wrote for predominantly male audiences, and Marcus Aurelius wrote only for himself. In texts aimed at men, being womanly or effeminate would, in fact, be a negative trait for Stoics, because it would not be in accordance with men's natures. For example, when Cicero says that there is nothing worse or more shameful than an effeminate man (*quid est autem nequius aut turpius effeminato viro?*), he does not necessarily also mean that it would be bad for a *woman* to be womanly (*Tusculan Disputations* 3.17.36). Nevertheless, it is far from clear that the Stoics universally considered womanliness to be a positive trait even for women to hold. Musonius still calls female virtue *andreia* and argues, "Someone might say that courage (ἀνδρεῖα) is an appropriate characteristic for men (ἀνδράσιν) only, but this is not so. It is also necessary for a woman—at least for a most noble one—to be courageous (ἀνδρίζεσθαι) and free from cowardice so that she is overcome neither by pain nor by fear" (lecture 3.4).[62] The predominance of words with the root *andr-* shows that Musonius believes it would be positive for a woman to be manly, even though it would never be positive for a man to be womanly. Manliness is an inherently positive trait; womanliness is not.[63]

The inherent sexism of Stoic thought continues to echo in the present day, even outside of far-right online communities. The neo-Stoic movement has been plagued by accusations of a persistent gender imbalance, which Massimo Pigliucci responded to in an interview on the blog of the American Philosophical Association:

Sometimes people say that Stoicism is more popular among men because it is about suppressing emotions, but that gets it twice wrong: first, because that's actually a profound mischaracterization of the philosophy; second, because it uncritically accepts the stereotype that women are more "emotional" (and therefore more fragile?) than men. I hope we are finally moving beyond that sort of false biological dichotomy.[64]

Pigliucci is correct about both Stoic philosophy (which emphasizes understanding emotional reactions, not suppressing them) and about the damaging nature of gendered stereotypes about rationality. But he does not account for the fact that there is a large community of men who are interested in Stoicism and who will never "move beyond" that biological dichotomy—men who believe, in fact, that their acceptance of that dichotomy is a sign of their enlightenment.

Stoic White Men

The Roman historian Sallust wrote of the famous Stoic Marcus Porcius Cato that he preferred to be good rather than to seem so (*Bellum Catilinae* 54.6; *esse quam videri bonus malebat*). For writers in the Red Pill, however, seeming to be Stoic and talking about Stoicism are at least as important as becoming a dedicated practitioner of the philosophy. Their writings often reveal that their interest does not lie in using Stoic precepts to become more virtuous. Instead, they project the *appearance* of emotional control by quoting Stoic writers in order to change their image from a group of angry white men to the only people brave enough to speak truth to power.

The Red Pill blogger who goes by the pseudonym Rollo Tomassi responds to the idea that the manosphere is full of angry men in the post "Anger Management" on his blog *The Rational Male*:

> But are we angry? I can't say that I haven't encountered a
> few guys on some forums and comment threads who I'd

characterize as angry judging from their comments or describing their situations. For the greater whole I'd say the manosphere is not angry, but the views we express don't align with a feminine-primary society. Men expressing a dissatisfaction with feminine-primacy, men coming together to make sense of it, sound angry to people who's [sic] sense of comfort comes from what the feminine imperative has conditioned them to. Most of the men who've expressed a genuine anger with me aren't angry with women, but rather they're angry with themselves for having been blind to the Game that they'd been a part of for so long in their blue-pill ignorance.

In addition to denying that anger is omnipresent in the manosphere, Tomassi also asserts that anger is unfairly gendered masculine: "Accusing a man of misogyny will always be more believable than accusing a woman of misandry because men are always just angrier than women."[65] This statement is arguably more revealing about his limited definition of misogyny, of which he seems to think anger is a key component, than it is an accurate assessment of how anger has been gendered.

Because Tomassi is convinced that the men of the manosphere are not angry (except at their earlier, blue-pill selves), he rejects sociologist Michael Kimmel's conclusions in his 2013 book *Angry White Men*, a study that argues men in the United States are suffering from what Kimmel calls "aggrieved entitlement," anger that something they believed they were owed—sex, money, power—has been stolen from them by women, people of color, and immigrants.[66] Kimmel never pretends to be disinterested; he tells one subject who shows up to their interview wearing a Confederate-flag shirt that his job is not to agree with the man's viewpoint, just to try and understand it.[67] In a 2014 post on *The Rational Male*, Tomassi dismisses Kimmel's entire project as soon as he sees the first word of the title and assesses that writer is a "beta male": "Because of this embrace of feminine-primacy, the Professor is probably not the best equipped to educate men on issues of anger. As such, my guess is he cannot discern the difference between aggression born from anger and aggression as a vetting and honing mechanism of the male psychology."[68]

Tomassi is not alone in this belief: the men who frequent Red Pill sites are confident that they are more rational and less emotional than anybody else. It is irrelevant that few outside their community agree that these men are reasonable, calm, emotionally controlled individuals. They consider themselves impervious to outside opinions as long as they maintain frame, as addressed in Chapter 1. Their concept of frame control—the idea that each person should craft a perception of the world, also called an "evaluative outlook" by Stoicism scholars, that cannot be shaken or shifted by outside opinion—has enabled them to continue believing that they are rational and unemotional instead of angry and abusive.

Tomassi bases his essay and the 2017 follow-up "The Anger Bias" on the assumption that anger is generally considered to be a male characteristic and is therefore stigmatized in order to further what he calls "the feminine imperative."[69] But anger, like all emotions, has also been gendered feminine. The angry feminist is as common a stereotype in contemporary society as the angry white man, and every feminist has been told many times that if she would just calm down and speak rationally, she might convince more people—a phenomenon sometimes called *tone policing*. The idea that women are inherently angrier than men goes back to the ancient world: Seneca says in *De Ira* that women and children are angrier than men are (1.20).[70] The Red Pill community has enthusiastically adopted this view. They believe that women are *by nature* more emotional and less rational than men—a position expounded at length in a two-part article on *Illimitable Men*, "The Myth of Female Rationality."[71] Ideas about female nature, often understood with reference to pseudo-evolutionary psychology, are then buttressed with support from Stoic philosophy.

The perpetuation of systemic racism is also coded into how Stoicism is understood in the Red Pill community, although perhaps more subtly than sexism is. In addition to being gendered, anger is also heavily racialized: the angry black man and angry black woman are both well-worn tropes.[72] It would be difficult to overstate how damaging these stereotypes have historically been for the black community. These familiar labels turn what might otherwise have been perceived as righteous indignation at deep systemic inequalities into irrational rage. They trivialize the emotions of a population that makes up more than ten percent of our country. They have measurable negative impacts on health and quality of healthcare, especially for mental health.[73] Worst of all, they reinforce the idea that black people, especially black men, are inherently angry and therefore more prone to violence. And if you believe, as the ancient Stoics did, that reason is what separates humans from animals, then labeling black people as characteristically angry suggests that they are somehow less human than more rational people are—such as, say, the Stoic white men of the Red Pill community.

Racism is at the forefront of the discourse in the Alt-Right, but even those who do not openly advocate for a "white ethno-state" still often perpetuate racist views. In *The Obstacle Is the Way*, Holiday praises the fortitude that middleweight boxer Rubin Carter showed when he was wrongly convicted of triple homicide and given three life sentences—"He would not wear a uniform, eat prison food, accept visitors, attend parole hearings, or work in the commissary to reduce his sentence"—and writes that "many great figures, from Nelson Mandela to Malcolm X" learned that an unjust prison sentence is a blessing in disguise.[74] Holiday does not address the fact that all of the men he mentions in the context of prison are black. He may even have thought that the inclusion of a few black men gave his book some diversity: only a tiny portion of the individuals he mentions who displayed Stoic virtues are people of color. Through his use of almost exclusively white male exemplars, Holiday implicitly assimilates Stoic virtue to white maleness. And in his pages-long encomium to Rubin Carter, he fails to mention that in our society, black men are imprisoned at a rate far above white men, to the point that, as Ta-Nehisi Coates writes, "For African Americans, unfreedom is the historical norm."[75] It is irrelevant to Holiday's self-improvement philosophy that a prison sentence is an obstacle that black men are much more likely to have to overcome than white men are.

The reception of Epictetus in the Red Pill community provides an especially revealing glimpse into their approach to race. Articles about Epictetus unfailingly mention that he was a slave, but not in such a way as to start a meaningful conversation about power, society, and philosophy. In an article called "Viewing Stoicism from the Right" in the alt-right online magazine *Radix Journal*, Charles Jansen mentions Epictetus's Stoic lack of reaction to having his leg broken by his master. Jansen then attempts to critique Stoicism by writing, "Maybe Epictetus could have prevented his master from breaking his leg if he tried to stir up some doubts in the master's mind by targeted questions," showing—as Holiday did—an absence of awareness that there would be negative consequences for a slave talking back to his master or a prison inmate refusing to wear his uniform.[76]

Instead, the lesson these men take from Epictetus is that slavery is not detrimental to philosophical greatness. The responsibility for self-improvement is shifted onto the slaves themselves, absolving slave owners (and those who have benefited from the legacy of slavery) from the need to make amends. These men do not support reparations, affirmative action, or anything else that would acknowledge or ameliorate our own country's legacy of slavery and its long-term ramifications for black Americans. Maybe if they had been more like Epictetus, Red Pill Stoics seem to imply, their oppression would bother them less. Then there would be no need for movements such as Black Lives Matter, which the manosphere considers a hate group.

There is some debate among Stoicism scholars over whether Stoics were comfortable with the institution of slavery only because they were not perfect sages yet. Concerning Stoics and slaves, David Engel writes,

> Stoics never say that slaves should stop being slaves and become philosophers, but that they should practice philosophy without ceasing to be slaves. If you are a slave, you should practice philosophy, the Stoics think. But they also think that philosophizing is perfectly compatible with your remaining a domestic drudge; the fact that your virtue is no different from that of your owner is, apparently, quite separable from the fact that you occupy a particular position in society. Being a Stoic does not mean having to give up your slaves, and being a Stoic slave does not mean being any less a slave in society's eyes.[77]

In theory, the Stoics believed in an ideal cosmopolis, a "world city" where everyone would be an equal citizen regardless of class, ethnicity, or gender.[78] In practice, it seems that many of the Stoic writers whose works we still have were highly conventional in their approach to social issues.

The only true requirement for being part of that cosmopolis was living in accordance with reason. As I mentioned, the Stoics believed that the universe was designed by an inherently rational organizing principle. Everything that occurs happens according to that rational design, so it would be irrational to reject the way things are. This is the standard interpretation of Zeno's much-discussed admonition that people should "live consistently"—to which later Stoics added "in accordance with nature."

But rationality may itself be a product of privilege, so access to the cosmopolis might not be equally available to all. Martha Nussbaum argues that a central failing of ancient Stoicism is "the failure to understand the extent to which human dignity and self-respect require support from the social world."[79] In other words, every person in every situation will *not* be equally capable of the kind of self-control and *apatheia* that allows one to achieve perfect virtue.[80]

But does Stoicism *necessarily* perpetuate inequality? Or is it merely a philosophy that invites appropriation by groups such as the Red Pill that have an ideological investment in inequality? Scholars are divided on this question. Some even dispute whether Stoicism and progressivism are by nature mutually exclusive. Philosophy scholars Scott Aikin and Emily McGill-Rutherford agree that Stoicism has an "uneven track record by feminist lights" on women's issues, but contend that this "is not evidence that Stoicism *must have* such a problem. As a consequence, a consistent and morally sound Stoic feminism is possible."[81] Martha Nussbaum argues that an essential part of Stoicism is "a commitment to treating similar cases similarly unless there is a morally relevant distinction between them," implying that class, race, and gender should not be important considerations.[82] According to this argument, although Stoic *proficientes* who have not yet reached the level of sagehood may exhibit misogynistic or racist views, for a true Stoic sage, virtue would be incompatible with discrimination, rendering a Stoic progressivism possible. Whether Stoic progressivism is *desirable* is another question altogether.

In Stoic thought, as in contemporary America, emotional control is a sign of moral superiority—so the shift from angry white men to Red Pill Stoics allows the men who tout the benefits of Stoicism to rhetorically establish their moral superiority over the groups they perceive as irrational, including women, people of diverse sexualities and genders, and people of color. To the Red Pill Stoic, the feelings of anger and frustration experienced by these disempowered communities are actually moral failings, proof of their lack of internal fortitude, moral compass, and right to hold authority in the public sphere. An impassioned dedication to social progress means that one has not yet achieved emotional equilibrium. Many of the most influential writers in the Red Pill share the immovable conviction that they are the only ones who are able to overcome their emotions and see the world rationally. Since the Stoics believed that rationality was a virtue, simply being a part of the Red Pill community becomes virtuous—in the technical, philosophical sense.

These men, in their stubborn denial of their own anger, have hit upon an essential truth about those whom they categorize as social justice warriors. They are not entirely wrong in their belief that many women, people of color, and queer people *are* angry. In the book *Killing Rage*, bell hooks writes that "many African Americans feel uncontrollable rage when we encounter white supremacist aggression. That rage is not pathological. It is an appropriate response to injustice."[83] Audre Lorde has argued that anger is in fact *necessary* to fuel the drive to create social change. In her 1981 address "The Uses of Anger: Women Responding to Racism," she posits that anger can be a force to create a truly intersectional feminism:

> Every woman has a well-stocked arsenal of anger potentially useful against those oppressions, personal and institutional, which brought that anger into being. Focused with precision it can become a powerful source of energy serving progress and change. And when I speak of change, I do not mean a simple switch of positions or a temporary lessening of tensions, nor the ability to smile or feel good. I am speaking of a basic and radical alteration in those assumptions underlying our lives. . . .

Anger is an appropriate reaction to racist attitudes, as is fury when the actions arising from those attitudes do not change. To those women here who fear the anger of women of Color more than their own unscrutinized racist attitudes, I ask: Is the anger of women of Color more threatening than the woman-hatred that tinges all aspects of our lives?[84]

This argument would seem to suggest that a Stoic intersectional feminism is impossible—indeed, that any Stoic activism is impossible. Engel likewise argues that feminism is *never* compatible with Stoicism, because the issues that are close to feminists' hearts—equal opportunity, equal pay, reproductive rights, minimizing sexual and domestic violence—would be considered indifferents to Stoics.[85] Simone de Beauvoir made a similar anti-Stoicism critique in *The Ethics of Ambiguity*: "If a door refuses to open, let us accept not opening it and there we are free. . . . But no one would dream of considering this gloomy passivity as the triumph of freedom."[86]

The Red Pill challenges the validity of this anger on two grounds. First, they believe that the people who are truly discriminated against in our society are white heterosexual men. They often discuss "female privilege"—particularly the "pussy pass," which allows for gender-based injustice such as the fact that men who hit women face far worse consequences than do women who hit men. However, they also assert that they are not *angry* about this discrimination, because they practice Stoicism. For the true Stoic, righteous fury is not a meaningful concept: anger is the result of a false perception that an injustice has been committed against the self, which results in a desire for retribution.[87] Angry activists are truly inferior to the more rational people who do not give assent to such false perceptions of injustice.

This activist critique of Stoicism—that anger is necessary to promote systemic change—is of long standing, and it has found purchase on both the Left and the Right. Not all men associated with the Red Pill have embraced Stoicism completely. A few pages ago, I analyzed a passage from "Viewing Stoicism from the Right" by the alt-right writer Charles Jansen. Ultimately, at the end of that piece, Jansen rejects Stoicism entirely, because he believes the Alt-Right should nurture its anger: "I don't want to renounce to [sic] rage: I want it to motivate actions that will bring true justice on the world. . . . Rage can bind us as brothers of fortune, it leads us towards sharing a sense of justice and sharpening a sense of beauty. I don't want happiness—I want victory, I want what defines me and us to shape the world." Within the context of the Alt-Right, Jansen's assessment of "what defines me and us" is likely a euphemism for white supremacy.[88]

Many Red Pill Stoics tacitly accept Jansen's premise that anger is an appropriate response to injustice. Quintus Curtius claims in *Thirty Seven* that he still allows the foolishness of others to bother him too much: "A major sign of progress is to maintain one's calm in the presence of the avalanche of nonsense which the world throws at us. I confess I need more improvement in this area."[89] Valizadeh talks his way around the passivity he sees in *Meditations*, the book he otherwise praised in a 2016 review as "perhaps the most important book I've ever read," writing, "One complaint I have is that Aurelius seems to be proto-Christian by adopting passivity when it comes to fighting enemies. . . . Then again, it is in my nature to fight back against those who harm me, and doing so doesn't directly conflict with Aurelius's teachings."[90] Valizadeh's interpretation of the Stoic principle of living consistently in accordance with nature is entirely dependent on what he perceives his own nature as an individual to be; the world must bend to him. All three writers emphasize, through denial of perfect Stoic *apatheia*, that they believe there is unfairness in the world that is worth getting upset about.

Activist Stoics, on the other hand, have argued that a true Stoic should look after the welfare of all who have a share in the *logos*. Struggling against injustice is therefore in no way incompatible with Stoicism, provided that the impetus for the struggle comes from a search for justice and not from one's *pathē*. Classicist Christopher Gill made this argument at STOICON 2016, where he said, "The ancient Stoics did urge us to accept, in a calm spirit, things that are genuinely inevitable—above all, the fact of our own future death and that of other people, including those close to us. But this does not mean that we should accept unjust situations, which are not inevitable and are the result of deliberate human action." If you agree with this interpretation of the philosophy, then both intersectional feminists and members of the Alt-Right can embrace Stoicism freely and without concern.

Since the men of the Red Pill are convinced of their own superiority, however, they have absolute conviction that white male supremacy is justice. They are more rational, and more Stoic, and therefore the world would be a better place if they were in charge. The Stoics believed that the wise should treat those who are ruled by their emotions as they would treat children. Do not feel compassion for them—compassion is not consistent with Stoicism, since it involves considering as one's own something that is outside oneself. Epictetus writes in *Enchiridion* that, if you see somebody grieving, "as far as words go . . . don't hesitate to sympathize with him, or even, if the occasion arises, to join in his lamentations; but take care that you don't also lament deep inside" (16). But comfort them as you would a child, and guide them on the right path. For the Red Pill Stoic, systemic change that restores patriarchy would benefit everyone: since women and people of color are irrational and need guidance, society would be better off if rational white men were placed in charge.

What is the best way to respond to Red Pill Stoicism? The strategy that is used most frequently, it seems, is to critique these writers for their superficial misreadings of Stoicism, as Pigliucci does. While this kind of approach can be satisfying—and it is particularly tempting to point out the radical attitudes the ancient Stoics espoused about gender equality—nitpicking the flaws and errors in Red Pill readings of Stoic texts is a fairly pointless project. The writers themselves will maintain frame rather than engaging meaningfully on the interpretation of ancient texts, and their audiences are not impressed by academic expertise.

Such an approach, ultimately, does little to reveal why it matters that Stoicism has become so popular in antifeminist online communities, and it fails to take into account that Stoicism could be a healthful and beneficial philosophy for men in the United States today. Many men, even outside the manosphere, seem to feel that while women are allowed or even encouraged to be emotional, men face a societal pressure to be stoic (that is, not Stoic, but rather to display no emotions) that can lead to an inability to deal with emotions in a healthy manner. That pressure could be a contributing factor to the much higher suicide rate among men than women in the Western world.[91] Stoicism's emphasis on self-reflection and therapy could even be an aid to men. Although a few ancient Stoics famously committed suicide, including Cato and Seneca, Stoicism is not a philosophy that encourages suicide unless one is forced into a situation where living virtuously becomes an impossibility.[92] Otherwise, Stoics believe that suicidal tendencies arise from hastily giving assent to a false impression that circumstances are bad when in reality they are merely indifferent. The adoption of true Stoicism could help remedy the problems caused by the social pressure on men to be stoic.

The most Stoic response, arguably, would be to accept that Red Pill Stoicism is—to paraphrase the Epictetus passage with which I began this chapter—something "not within my power" (*allotrion*) and to realize that considering it my own (*idion*) will only lead to becoming "disturbed." Even if these men are weaponizing ancient literature and philosophy not to become more virtuous but to gaslight women, I should accept that their discourse is their responsibility, not mine. On the other hand, if I perceive that the Red Pill is promoting injustice in the world, the rational choice might be to attempt to lessen the impact of that injustice.

It will require heroic effort for the leaders of the neo-Stoic movement to ensure that, by embracing Stoicism, they are not perpetuating systemic oppression along the axes of gender, race, and class. Doing so will require more than simply pointing out that not all dead white men were openly misogynistic and that prominent Stoics advocated for gender equality and female education and against the sexual exploitation of women and slaves. It will require continued interrogation of how Stoicism helps both the oppressed (by helping people cope with their situation and make the best of it) and the oppressors (by drawing attention away from systemic injustice). Most of all, it will require us to confront why the works of writers such as Marcus Aurelius have proven so attractive to men like Valizadeh, Quintus Curtius, and Holiday—and what harm ancient literature can do in their hands.

3

THE OVID METHOD

The introduction to Julia Hejduk's recent translation of Ovid's *Ars Amatoria* opens with this declaration:

> Ovid's *Ars Amatoria* (Art of Love) *has one of the funniest premises of any work of literature*: namely, that Love—by which he means the initiation and maintenance of sexual relationships—is a field of study, like chess or astronomy or agriculture, whose strategies can be analyzed and taught.[1]

The *Ars Amatoria* is from the earlier part of Ovid's career, when he focused almost exclusively on erotic works.[2] It is an instruction manual for seduction in three volumes: the first teaches its readers how to seduce a woman; the second focuses on how to keep her interest; and the third teaches women how to seduce men. A follow-up volume, published a few years later, teaches men how to fall out of love with unattainable women.

Hejduk's assertion represents the general consensus among scholars about the *Ars Amatoria*: it is nearly impossible to find an article or book about the poem that does not contain words such as *amusing* or *ludic* (from the Latin *ludus*, "game"). By awkwardly forcing a poem about seduction techniques into the genre of didactic poetry—a genre mostly used for long technical treatises on subjects such as farming or ethics—Ovid is playing a complex poetic game with the reader's expectations. His use of the elegiac couplet, the meter of erotic poetry, instead of the dactylic hexameter, the meter of epic and "traditional" didactic poetry, emphasizes the disconnect between the poem's form and its content. Imagine a college textbook about the history of cupcakes, and you have the idea.

But the classicists who find the poem's premise so clever may not be aware of how seriously that same premise is taken by the pickup artist (PUA), or "game," community. A pickup artist is an individual, usually a heterosexual man, who has intensively studied and attempted to master techniques to convince women to have sex with him; these techniques fall on a spectrum from flirting to manipulation to harassment to assault. A few influential members of the community run blogs or publish books to disseminate their knowledge. To paraphrase Hejduk's claim about love in Ovid, pickup artists believe that "the initiation and maintenance of sexual relationships" are "field[s] of study . . . whose strategies can be analyzed and taught."[3]

The relevance of Ovid to the game community goes beyond a similar, strategic approach to seduction. In an attempt to give themselves legitimacy and gravitas, some pickup artists look back to famous seducers from history and reposition them as the intellectual predecessors of the modern seduction community—and Ovid is one such venerated figure. Neil "Style" Strauss, in his 2005 memoir *The Game: Penetrating the Secret Society of Pickup Artists*, writes,

Sure, there is Ovid, the Roman poet who wrote *The Art of Love*; Don Juan, the mythical womanizer based on the exploits of various Spanish noblemen; the Duke de Lazun, the legendary French rake who died on the guillotine; and Casanova, who detailed his hundred-plus conquests in four thousand pages of memoirs. But the undisputed father of modern seduction is Ross Jeffries, a tall, skinny, porous-faced self-proclaimed nerd from Marina Del Rey, California.[4]

Strauss is not alone in naming Ovid the ancient father of seduction. The *Ars Amatoria* is widely accepted by the community as the starting point of the teaching of seduction. Ovid's name is casually mentioned not only in Strauss's memoir, but also in blog posts across the seduction blogosphere with titles such as "The History of Pickup and Seduction, Part I" and "Recommended Great Books for Aspiring Womanizers."[5]

The reason for Ovid's inclusion in these lists is clear. Ovid gives men advice for how to seduce women, and pickup artists are men who define themselves by their interest in seducing women, so Ovid was arguably one of their own. But the concepts underlying how we think about sex and sexual relationships today—our ideas about sexuality, gender, race, and class—have shifted considerably from the conceptual categories that existed in first-century Rome. Both Ovid and modern pickup artists may be concerned with how to successfully obtain casual sex, but what would *casual* even mean in a Roman context? If one defines sex as casual if it is enjoyable but without serious consequences, then in a world where adultery was criminal, birth control was less foolproof, and abortion and childbirth were both life-threatening, sex could never be truly casual.[6] Ovid's advice necessarily puts women at risk.

The category differences are especially stark if, as Ovid later claimed and as many scholars now believe, the Ovidian *puella* (*girl*, or in PUA terminology, *target*) is a *meretrix* (*Tristia* 2.303–304).[7] A *meretrix* was an expensive sex worker, the sort for whom scholars often use the old-fashioned word *courtesan*, although the modern category it maps onto most closely might be the "sugar baby." The livelihood of the *meretrix* depended on leveraging her sexuality into financial security. So successful deployment of Ovid's advice would be more than physically dangerous to the woman; convincing a *meretrix* to enter into a sexual relationship for free would put her in a fiscally precarious position. It could create legal challenges for the man as well. Despite Ovid's early claim that he only promotes sex that is within the boundaries of the law ("there will be no crime in my song," *inque meo nullum carmine crimen erit*), the poem subverts the *leges Iuliae*, the moral legislation Augustus introduced that punished men and women for both adultery and for remaining unmarried for prolonged periods of time (*Ars* 1.34). As I will discuss later, Ovid's transgressions against these laws in the *Ars Amatoria* would eventually have grave consequences for him.

Even though it was published two thousand years ago, the *Ars Amatoria* can still feel very relevant to today's world.[8] But it is ultimately a poem that Ovid intended for *his* time, not for *all* time.[9] The superficial similarity between his suggestions for how to avoid buying your *puella* expensive gifts and advice on seduction blogs for how not to buy a girl drinks or spend more than twenty dollars on a date is misleading. Treating Ovid as "the original PUA"—or claiming, as Strauss did on Reddit, that "What works has always been the same throughout history, from Ovid's writing on seduction to today"—is difficult to justify from a theoretical perspective.[10] Although there are undeniable similarities, most are superficial, and the cultural conditions that shaped Ovid's text are entirely different from those that shape the seduction community.

To understand the seduction community, one must first discard the common misconception that "having game" means only that a man is skilled and experienced at seducing women. Seducing women is a necessary condition of having game, but it is far from the entire story. The truth is more complex and less teleological. Game is about becoming the kind of man women will be attracted to automatically. Such a man—stylish, confident, charismatic—will, as a matter of course, receive everything he wants, including desirable sexual partners. The seduction community claims it can teach almost anyone how to become that kind of man.

Game is about more than learning techniques for how to pick up women in bars; it is a way of thinking about women, men, and how the two genders interact with each other in social settings. This rethinking of the rules of attraction makes pickup artist ideology simultaneously seductive to men and dangerous for both men and women. If game were, as its proponents sometimes claim, simply about helping men gain confidence in their interactions with women, then it would be relatively harmless. It could even be a positive force to help men overcome social anxiety. But that is not really what having game is about—and, in fact, many members of the manosphere mock that way of thinking because it places the power in the hands of women, who are permitted to decide whether a man is attractive or not. At its heart, learning how to have game requires conceptualizing and internalizing ideas about gender that lead to devaluing women and relegating them to the status of sexual objects. Even worse, many leaders of the seduction community believe that women are *designed* to be sexual objects (and, later, mothers) and that any woman who cares about her education and career—or any woman who does not need validation from men—is unnatural and perverted.

The seduction community's claim that Ovid is their predecessor fits into the general philosophy of masculinity that pervades websites such as *Return of Kings*. These sites want to portray their interpretation of masculinity, for which they have coined the term *neomasculinity*, as revolutionary and subversive, but with strong historical roots and a basis in biological fact—or at least what they perceive as biological fact.[11] Masculinity and femininity, imagined in this way, are not social constructs. They are concepts with a fixed, ahistorical, essential meaning from which we have deviated over time but to which we can return.

If you think about gender expression in this way, comparing Ovid to modern game guides presents few methodological challenges. According to this reading, the categories *man* and *woman* have remained static over the past two thousand years. If one believes that the female members of the human species are programmed by nature rather than conditioned by culture to respond to certain kinds of male behavior, then Ovid's suggestions can certainly still have merit. Having game is about being the kind of man women want, and pickup artists believe that women have always wanted and will always want the same thing: an "alpha male" to provide for them and father their children.[12] This idea is perhaps most clearly expressed in Peter Burns's 2015 article on *Return of Kings*, "Lessons from PUA Ovid: The Original Latin Lover." Near the end, Burns argues,

> The "Ars Amatoria" is quite red pill. Ovid frequently mentions cases of female promiscuity and how many women are hypergamous. Female nature does not change according to the country or the age. Women of two thousand years ago and from very conservative cultures had the same vices that the women of today have.[13]

It does not matter to pickup artists that the science underlying this thinking has been repeatedly debunked. Since the invention of the Kinsey scale, most educated people have accepted that sexuality falls roughly along a spectrum. In recent years, gender expression from masculine to feminine has also been conceptualized as a spectrum rather than a strict binary. Scientists now think that, along with sexuality and gender, biological sex is not as binary as previously thought.[14] The Intersex Society of North America recognizes more than a dozen intersex conditions that challenge the strict definitions used by the manosphere: there are individuals with ambiguous or multiple sets of genitalia, people whose chromosomal makeup does not match their genitalia or sexual organs, men and women with hormonal irregularities that lead them to present as unusually feminine or masculine. A seventy-year-old father of four was recently discovered to have a uterus.[15] Such nuances are lost on the writers of the manosphere, most of whom can only comprehend one transhistorical model of manhood and womanhood. The same fallacious thinking about what it means to be "male" or "female" underlies both their gender essentialism and their use of Ovid.

A close analysis reveals the weaknesses and inaccuracies in the seduction community's use of Ovid's *Ars Amatoria*. But their reading of the text also poses a fundamental and very important challenge to how the poem is usually interpreted. The fact that pickup artists have so closely embraced this text is reason to look more closely at its disconcerting aspects.

Several times over the course of the poem, Ovid's narrator advises his reader to act in a way that today would certainly be considered sexual assault. The *Ars* is not Ovid's only text that depicts rape in such a disturbing manner; as I mentioned in Chapter 1, Ovid's masterpiece *Metamorphoses* has also come under fire for its depiction of sexual violence. One scholar even reports one of her students calling *Metamorphoses* a "handbook on rape."[16] While this description fits *Metamorphoses* somewhat uncomfortably—the text has many stories about rape, but is hardly an instruction manual—one could certainly argue that "handbook on rape" is an accurate description of the *Ars Amatoria*. Treating the premise of this poem as fundamentally playful or subversive, as some scholars do, becomes irresponsible when there is a community using it today to normalize an attitude toward consent that would not be out of place in ancient Rome. The fascinating and the unnerving aspects of Ovid's work cannot be detached from each other. And Ovid's funny, charming, disturbing pickup-artist manual shows that manipulating and abusing women is an integral part of seduction technique.

Authorship and Authority in Seduction Manuals

The seduction community is a fully developed ecosystem with complex social dynamics, and the kind of people who rise to prominence within the community can tell us a great deal about what the community itself values. Nearly every successful purveyor of seduction advice uses the same narrative: when he started out, he was completely unsuccessful with the opposite sex, but he gradually learned specific, easily replicable techniques to gain female interest, and then decided to give other unfortunate men the advantage of his hard-earned knowledge.

In her book *Confessions of a Pickup Artist Chaser: Long Interviews with Hideous Men*, feminist sex writer Clarisse Thorn creates a taxonomy for classifying different kinds of pickup artists. She groups them into six categories:

1) The Analysts, who use seduction methodology "to help them understand sex and gender."
2) The Freaks and Geeks, "guys so shy and awkward that they've never had a single date."
3) The Hedonists, who are looking to have as much sex as possible.
4) The Leaders, community organizers who want to teach other men.
5) The Sharks, looking to monetize the community.
6) The Darth Vaders, misogynists who "feel resentment, contempt, and distrust for women."[17]

From her research, Thorn estimates that most, perhaps as many as eighty percent, of the hundreds of thousands of men with accounts on game websites fall into the second category, freaks and geeks. Many of these men begin with little experience with the opposite sex, and once dating dynamics are explained to them and they are able to successfully find romantic partners, they stop participating in the community.[18] Seduction is a tool for such men, not a lifestyle. The men writing about how Ovid is the original pickup artist are not in this group; they usually exhibit qualities from the analyst and leader categories, although some are certainly "Darth Vaders" as well.

Many members of the community fit several of Thorn's archetypes, either at the same time (leaders are often also sharks) or over the course of a period of time (a geek who becomes skilled at game can become a hedonist and then a leader). It may be more helpful to think of Thorn's classification system not as a group of discrete categories, but as a set of characteristics that pickup artists exhibit in different proportions. I would classify Ovid's narrator as mostly an analyst, although he embodies elements of all six categories.

The strategies used by the community's leaders to establish their authority are similar to those used by Ovid to construct his narrator's authority in the *Ars Amatoria*. Nick Krauser, promoting his ninety-eight dollar textbook *Daygame Mastery*, writes, "There are *specific actionable foolproof steps* you can take to become the naturally charismatic sexworthy man."[19] An instruction manual for obtaining natural charisma may seem oxymoronic, but it perfectly encapsulates the strange and self-contradictory business of establishing oneself as a professor of love (*praeceptor amoris*). First, the would-be *praeceptor*, ancient or modern, must convince his reader that love is both learnable and teachable; then he must explain what makes him the best possible teacher.

The writers of these texts preach that looks, money, and charisma are not requirements for seducing beautiful women; the only true requirement is technique. Ovid's *praeceptor*—the narrator of the *Ars Amatoria*, loosely but not perfectly identified with Ovid himself—compares *amor* to skill at physical activities: "Art (and sails and oars) is what makes speedy ships move, art drives light chariots: art is the thing to steer Love" (*Ars* 1.3–4). He even compares himself to Daedalus, the paradigmatic craftsman of myth: as Daedalus created wings for Icarus, Ovid will master Cupid, the winged god (*Ars* 2.21–98).[20] Ovid exhibits complete confidence that no woman is unattainable if you use the right seduction techniques on her:

> First, you've got to believe in your heart that they all can
>   be caught:
>   you'll catch them, you just need to lay your trap!
> Birds will sooner keep quiet in spring, cicadas in summer,
>   the Maenálian hound
>   sooner show his back to the hare,
> than a woman will fight back when assailed by a young
>   man's flattery;
>   even the one you might think doesn't want it will want
>   it. (*Ars* 1.269–274)

The self-evident danger of this view is that it encourages the reader to think of female consent as a foregone conclusion once the pickup artist becomes sufficiently skilled.

The most common English translation of *Ars Amatoria*, *The Art of Love*, masks the text's true nature and makes it sound much more flowery than it really is.[21] For one, the Latin term *amor* does not map neatly onto the English word *love*; it has connotations of desire and sex that *love* does not always have. Regardless, *Amatoria* does not even mean "of love." That would be *amoris*, or perhaps *amandi*. It means "of the *amator*," the lover.[22] For another, the word *ars* does not mean art as we usually refer to the arts—it means a skill or technique, like training a horse or stone masonry, as the Greek word *technē* does. *Ars* is often contrasted in Latin with *ingenium*, which means something like natural talent (literally, the *genius* that is inside of you). A better translation of the title *Ars Amatoria* might be *How to Become an Expert Lover.*

The Mystery Method, originally published in 2005 as *The Venusian Arts Handbook*, is a foundational text of the modern seduction community. Although the techniques taught by Erik "Mystery" von Markovik are considered somewhat outdated now by the game community—and perhaps even immoral—von Markovik created much of the terminology and jargon that is still used today.[23] (See the Glossary for a list of some of these terms and acronyms.) His framing of pickup as falling into three phases—"attract, comfort, seduce"—remains influential.[24] The Mystery Method relies heavily on understanding and utilizing the interplay between power, gender, and sex. Von Markovik thinks of sexual attraction as being fundamentally about power, although he uses the term *value.*

A successful pickup will ensure that the man has power over the woman through a dual set of strategies meant to convince the woman to lower her perception of her own value as her perception of his value increases. An approach begins with the man in a position of lesser power, because he already feels attracted to the woman and wants to sleep with her. She therefore has power over him. The Venusian / pickup artist attempts to reverse the power differential by showing his own value through "social proof," such as being famous or having a group of friends and wingmen around to prove his popularity; telling great and entertaining stories which, if memorized in advance and used repeatedly, qualify as "canned material"; and displaying little or no interest in the target. This lack of interest is performed through throwaway backhanded comments known as *negs*, meant to emphasize that he does not consider his target attractive, such as "Are you always this annoying?" or "Your nose wiggles when you laugh!"[25] Negs and "demonstrations of high value"

(DHVs), such as telling the target about one's professional success or exceptional skills, must be balanced with indicators of interest (IOIs) to keep her from feeling that she is being insulted and patronized. Von Markovik's basic underlying assumption is that women are programmed to seek male approval. Withhold that approval, and they will work twice as hard to get it.

In a way, having game means precisely what it sounds like—interactions between the sexes become a game and therefore winnable. Successfully picking up a woman becomes conceptually similar to a boss battle at the end of a level in a video game: if you have accrued the right armor and weapons—cultivation of personal appearance, canned material such as practiced opening lines and entertaining stories—and you have practiced enough that your technique and timing are perfect, almost any boss can be defeated. Or, to put it in more technical terms:

> The PUA world applies algorithms, testing and feedback, and gamification to human interaction, turning women into not just sexual objects but essentially treating that cisgendered biological configuration as a Turing-complete machine in which specifying the right sequence of inputs results in access to specific ports and protocols. In this case, the vagina is typically the desired port, with other orifices of interest as well. Working one's way through various handshaking protocols and debugging the process of obtaining port access is part of the game.[26]

This "gamification" is present in Ovid as well, in his emphasis on seduction methodology and his frequent use of comparisons to mythical heroes. Love is not an emotion, it is an activity—one that, if mastered, will make you the next Theseus or Hercules.

One problem with this gamified mindset, however, is that easy games are usually less fun for the players. A woman who is initially enthusiastic about having casual sex with you is less desirable than one who has to be coerced through proper seduction methodology. Von Markovik even dismissively refers to instances when a woman consents to sex without having seduction techniques used on her as "fool's mate," referring to a quick chess victory made possible by the weakness of one's opponent rather than through one's own skill.[27]

Not every man needs carefully honed techniques to find sexual partners. In *The Game*, Strauss distinguishes between "naturals," men seemingly born with the ability to attract women without trying, and "unnaturals," men who need to perfect the craft of seduction. He theorizes that the former had their first sexual experiences at such a young age that they never developed any insecurities with the opposite sex.[28] Tucker Max, the author of *I Hope They Serve Beer in Hell*, is precisely this type of natural "alpha male of the group" (AMOG), while Strauss is a self-identified unnatural. Max is in the minority among seduction teachers. Most, like Strauss, started off nervous and awkward around women, but they believe that, with enough practice and the right toolkit, anybody can learn how to be successful with the opposite sex.[29]

This is a seductive idea. Age, race, physical appearance, profession, wealth, education—although having the "right" attributes in these categories can help attract women, none of them is absolutely required. In fact, if your goal is not to seduce women but to teach other men how to seduce women, your money, natural charm, and good looks can actually be *detrimental* to your credibility, because your success can be attributed to innate attractiveness rather than the sharpness of your skills. Ross Jeffries—the "father of modern seduction" whom Neil Strauss describes as "a tall, skinny, porous-faced self-proclaimed nerd"—criticizes another instructor, David DeAngelo, by saying, "The guy is so fucking good-looking and well-connected in the nightclub scene it just astounds me that people think he could ever understand their situation and the difficulties they encounter in dealing with women."[30]

Strauss is not much more positive about his own appearance than he is about Jeffries's. He describes himself as "far from attractive," then proceeds to criticize his large nose, thinning hair, and "beady" eyes. He concludes, "When I look down at my pale, slouched body, I wonder why any woman would want to sleep next to it, let alone embrace it. So, for me, meeting girls takes work."[31] This seemingly negative self-assessment is really conceit masquerading as self-deprecation. The uglier you claim to be, the more impressive your carefully honed seduction skills will seem.

Ovid's own brand of "humblebragging" is not about his appearance, but about his poverty.[32] Instead of talking about his unattractive appearance or awkwardness, Ovid is ostentatious about his lack of wealth. He specifically excludes the rich as a potential audience for his book:

> I'm not making my way as a teacher of loving for the rich;
> whoever is willing to spend has no need for my art.
> He who says "Take!" when he pleases has got all the
> genius he needs;
> I give up—he's got more charm than my inventions.
> I'm a prophet for paupers, because I loved as a pauper;
> since I wasn't able to give gifts, I gave words.
> (*Ars* 2.161–166)

Ovid came from a wealthy family, and he was associated with several wealthy patrons of the arts, so it is doubtful that he himself was truly poor, although the poet and the narrator should not be identified too closely. The emphasis in this passage seems, rather, to be on *comparative* wealth: there are always men who are wealthier than you are, whom the object of your affection may prefer over you. There is widespread agreement among pickup artists—in fact, among all members of the Red Pill—that women are naturally "hypergamous," which is to say that they are always on the lookout for a wealthier and more successful man to trade up to. Aspiring pickup artists such as Peter Burns, whom I quoted earlier, have read Ovid's self-assessment as a boast that he was so successful with women that he was able to overcome their hypergamous natures and convince them to choose him over wealthier men.

Other negative attributes can also help an aspiring teacher of seduction establish his authority for potential students. In *The Game*, von Markovik, Jeffries, and Strauss have low-status jobs as, respectively, a magician, a hypnotist, and a freelance writer—but those professions are assets within the seduction community, because they do not afford the advantages that a high-status, high-paying job would have.

Among modern pickup artists, being a person of color might be another handicap with women that can be an asset in asserting credibility within the community. The biological essentialism of Red Pill ideology applies not only to gender but also to race, and pickup artists display stereotypical assumptions about the testosterone levels, penis sizes, and personality types exhibited by men of various races. It is a truism throughout the Red Pill community that alpha males get the most women, and they often claim that Asian men do not have enough testosterone to be alphas.[33] Black men, on the other hand, are believed to be natural alphas, but may have a more difficult time convincing high-value white women to enter into long-term relationships.[34] Overcoming either of these deficits would cement a man's position as a master of seduction.

Another unexpected asset for gaining authority within the game community is a history of rejection. Pickup artists have terms for men who have no game, such as *beta male*.[35] A term that shows up frequently in older seduction manuals is *average frustrated chump* (AFC). When trying to impress other members of the seduction community, pickup artists often boast about their previous awkwardness with women and the fact that they did not lose their virginity until they were in their twenties. Krauser's memoir nods to his previous beta days with its title *Balls Deep: From Jilted Lover to Lady Killer.* Roosh V claims on the promotional page for his *Bang* series of seduction guides that in college, one of his friends teased him that he would have been unable to "get laid in a whorehouse."[36]

Ovid presents a similar narrative arc to his readers, in which his earlier poetic works give him authority to present himself in the *Ars Amatoria* as the *praeceptor amoris*. But the *praeceptor* is not precisely Ovid; he is a persona. Therefore, his goals and opinions may not be those of the author. The *praeceptor* is, first and foremost, a teacher of seduction, but Ovid is a poet, and his motivations are harder to access.

Although they are not identical, Ovid and his *praeceptor* do share a biography—or, more precisely, a bibliography. Ovid's narrator repeatedly instructs his reader to acquaint himself with Ovid's earlier love poetry. The *praeceptor* uses Ovid's previously published texts to build his authority as a teacher, while Ovid's wit and turn of phrase establish his skill as a poet. The *Amores*, Ovid's first published work, are three books of love elegies in which Ovid often bemoans his lack of control over Corinna, his *domina* (mistress). If Ovid sometimes looks pathetic in these works, then that is all the better, because those experiences will help him relate to his audience in the *Ars*. Next in the Ovidian corpus were the *Heroides*, love letters in elegiac couplets written in the personas of various women from myth to their faithless lovers. With experience as a beta male from the *Amores* and insight into the female psyche from writing in their voices in the *Heroides*, the Ovidian narrator of the *Ars* is established in his credentials as a teacher of seduction.

The success of this self-fashioning is debated by scholars, many of whom see Ovid's *praeceptor amoris* as remarkably ineffective. He frequently contradicts himself and loses control over his own metaphors and lessons, and there is serious doubt as to whether the specifics of his advice would really work outside the world of elegiac poetry.[37] Pickup-artist readers of the text generally ignore the ironizing element of the narratorial voice in the *Ars* and instead read the text as straightforward advice. Likewise, non-pickup-artist readers of modern seduction manuals may find themselves suspecting that some of the suggestions for DHVs and negs are not meant entirely in earnest. But no matter how ironic the master of seduction's posture appears, the underlying theory he espouses nevertheless hints at deeply held assumptions about male and female behavior.

Apples and Oranges?

Strauss's statement that "what works has always been the same throughout history, from Ovid's writing on seduction to today," while an exaggeration, is among the *least* problematic pickup-artist analyses of Ovid.[38] Considering the temporal and spatial distance between Ovid and more recent *praeceptores amoris* such as von Markovik, there are indeed a surprising number of similarities in their methodologies, even if they do not match precisely. This has not gone unnoticed by many within the community, including Valizadeh.

Valizadeh is the author of seventeen books, including several game guides such as *Day Bang, 30 Bangs*, and *Bang Poland*. Some of these books have been criticized for their depiction of sex in which the women were so intoxicated that their ability to legally give consent is in question. Valizadeh is well aware of these criticisms and even seems to invite them with his writing; one passage from *Bang Iceland* reads, "While walking to my place, I realized how drunk she was. In America, having sex with her would have been rape, since she couldn't legally give her consent. It didn't help matters that I was relatively sober, but I can't say I cared or even hesitated. . . . If a girl is willing to walk home with me, she's going to get the dick no matter how much she has drunk."[39] In *Bang Ukraine* he describes physically holding a woman down during sex "to prevent her from escaping."[40]

Valizadeh mentions Ovid on occasion, and in 2011 he wrote a book review of the *Ars Amatoria* on his personal website.[41] He signals his intent to use the ancient author to legitimize his own reactionary gender politics at the beginning of the review, with the *Ars* quotation he chooses as an epigraph: "Women like being hurt. What they like to give, they love to be robbed of." This passage is taken from the part of Ovid's text that has most troubled scholars—the section in which the poet seems to recommend rape. Of the text as a whole, Valizadeh writes,

> I couldn't help but read what is arguably the oldest game book is [sic] existence, written by Ovid around 2 CE. Ovid teaches you how to be a gentleman who understand's [sic] what turns women on. While a lot of his advice is meant for a time where chivalry was rewarded, it's not surprising to see that many of his lessons still hold true today. . . . This book could also be called Bang Roman Empire for its specific venue advice on where to find women.[42]

In judging the *Ars* worthy to be titled like his own game guides, Valizadeh implies that his own works are the standard by which Ovid's manual should be judged. It is an odd and revealing choice.

The weaknesses of this interpretive strategy become clearer in the work of a small manosphere site called *Purple Motes*, a history blog run by Douglas Galbi, an economist trained at M.I.T. who has also contributed to the men's rights activism site *A Voice for Men*. In "Understanding Ovid's Satirical Roman Love Elegy," Galbi poses this rhetorical question about the *Ars Amatoria*: "Ovid was highly intelligent, extremely learned, and extraordinarily gifted poetically. Why did he create these literary outrages?" His response: "Prolific bloggers Roosh, Obsidian, and Roissy provide learned insight into

Ovid's love elegies." After describing the styles of these three blog-gers, whom he collectively terms "ROR," he writes, "If you want to understand Ovid's love elegies, hear ROR roar."[43]

In his discussion of Ovid, Galbi cites classicist Sharon James's 2003 monograph *Learned Girls and Male Persuasion: Gender and Reading in Roman Love Elegy*. When James emailed him to correct a mischaracterization of her stance, he responded publicly in a follow-up post, "More on Ovid and Roman Love Elegy," telling her that she had failed to take Ovid's male perspective seriously and re-iterating that she might understand Ovid better if she acquainted herself with the writings of the manosphere's new Ovids.[44]

Since Galbi venerates Roosh V, it is not a surprise that his view is the natural progression of Valizadeh's assumption that Ovid's work may deserve the title *Bang Roman Empire*. While Valizadeh implies that Ovid's *Ars* can be judged to have value only because it resem-bles his methodology in his own game guides, Galbi argues that Ovid's position cannot even be properly comprehended without reading and understanding the works of Valizadeh—and Mumia "Obsidian" Ali and James "Roissy" Weidmann—first.[45] Ovid is not only *similar* to modern pickup artists—he is one of them, and at-tempting to understand his literary aims without a proper appre-ciation of Ovid's pickup artist agenda will only lead to, as Galbi writes, "completely fail[ing] to understand central aspects of the male position."[46]

This idea has precedents in scholarship about reception studies and comparatism. Reception theory holds that the reader con-structs the meaning of a text in the moment of reading. Since the reader's cultural context necessarily influences the frame that he or she brings to a given reading, a later text can be said theoreti-cally to have "influenced" how we read even a much earlier text.[47] Similarly, Friedrich Nietzsche suggests in "We Classicists" that it is wrong to compare the ancient world to the modern for the pur-pose "of proving that what is valued highly in our times was valued by the ancients. The right starting point is the reverse: that is, to proceed from the recognition of modern perversity and then to look backward—many very shocking things in the ancient world then appear as profoundly necessary."[48]

Galbi argues that reading Ovid alongside today's modern *praeceptores amoris* can reveal Ovid's motivations in writing the text.[49] And it is true that the similarities in authorial posturing and seduction methodology suggest that—contrary to what many scholars believe—the *Ars Amatoria* might be a practical manual and not just a sophisticated poetic game. However, Ovid's text can also teach us how to be better readers of modern seduction manuals. The similarities suggest that Ovid and today's pickup artists shared some similar ideas about female agency, desires, and personhood—and that all of them believe that, on some level, all women are *meretrices*. Reading game advice and the *Ars Amatoria* alongside each other often reveals disturbing subtext lying just beneath the surface of apparently innocuous texts.

———

Since my intent is not to put together a complete catalogue of all the ways in which Ovid resembles modern game guides, what follows is not an exhaustive list of similarities. Game itself is too diverse and multifaceted, and Ovid's social context was too different. A detailed, side-by-side comparison of techniques would be the sort of project one might find on pickup artist websites, because their goal is to mine Ovid's text for useful advice and resonances with their own techniques.[50] I am less interested in comparison itself than in comparatively analyzing the *theory* underlying Ovid's text and modern game principles. In particular, the similarities between the ancient and modern seduction guides highlight two parallel tendencies: the prizing of male subjectivity over female subjectivity, and a project of gradually intensifying the violation of women's boundaries.

Some of the similarities between the advice in the *Ars Amatoria* and today's seduction manuals are relatively harmless, even obvious. For example, Ovid's *praeceptor* tells readers that, while they do not need to be exceptionally handsome to have success with women, it helps to be well-groomed and to dress in clothes that fit well (*Ars* 1.505–524).[51] More important than appearance is confidence. Play to your strengths, or, as von Markovik constantly tells his reader, "display high value." Ovid advises the seducer-in-training to find an excuse to show off his assets, be they his looks, singing, or drinking (*Ars* 2.497–506). He also has a clear sense of what pickup artists call "inner game," the self-improvement that will naturally lead women to find you attractive: "To be loved, be lovable / Something a handsome face alone won't give you" (*Ars* 2.107–108). To illustrate this principle, he tells the reader that Ulysses was able to hold Calypso's attention through his intelligence and storytelling skills (*Ars* 2.123–142).

Other similarities in advice, however, reveal more disturbing undercurrents. For example, a subset of game literature is dedicated to "day game," the art of picking up women on the street, in coffee shops, or anywhere else during daylight hours. Ovid, similarly, advises his reader on which of Augustus's favored monuments are most conducive to locating attractive women (*Ars* 1.67–90). In late August 2016, a minor controversy arose online around a day-game article on *The Modern Man* with advice for how to approach and strike up a conversation with a girl who is wearing headphones. Several feminists argued on Twitter and in think pieces that the advice was creepy and invasive: since headphones are often intended as a signal that a woman *does not* want to be approached, the underlying message of the original article was that men should ignore the personal boundaries of women they find attractive. Others thought that the original piece was relatively innocuous and not worth making a fuss over. But a comparison with the *Ars Amatoria* can show what, precisely, the feminist writers were concerned about:

She fears what she asks, she wants what she doesn't ask:
    your persistence.
Press on, and eventually you'll get what you want.
Meanwhile, if she's being carried along reclined on a
    couch,
    with a cunning pretense approach your mistress's
    litter;
and so no one will lend disagreeable ears to your words,
    cloak them slyly as you can in ambiguous signs.
Or if her feet are treading a spacious colonnade
    in a leisurely way, here again, make a friendly detour,
and be sure to go in front sometimes, sometimes fall
    behind,
    sometimes hurry along, sometimes go slowly.
And you mustn't be ashamed to drift a bit past the center
    of the columns, or to nudge her side with your side.
    (*Ars* 1.485–496)

At first glance, Ovid's narrator's advice also seems innocuous: a man approaching a woman being carried in a litter is hardly more invasive than a man approaching a woman wearing headphones. But both texts subtly argue that women do not have the right to travel from one place to another without being approached by desirous men, and both urge their readers to ignore hints that women may not be receptive to such approaches.

This violation of boundaries also extends to infringing on women's personal space—and, crucially, never asking for permission. In physical interactions with women, Ovid's advice is to take the lead. He provides suggestions for finding ways to touch your target, including brushing dust, even imaginary dust, off her at the racetrack or bumping into her in a crowd (*Ars* 1.135–156, 2.209–214).[52] Pickup artists call this sort of strategy to use physical contact *kino* (kinesthetic) *escalation*, techniques to encourage and provide opportunities to touch a target.[53] According to most game advice, you should escalate constantly, starting with relatively innocent touching and then gradually intensifying the physical contact and forcing her to signal to you when she wants you to back off by pulling away. Weidmann's thirteenth "Commandment of Poon" is to "err on the side of too much boldness." He writes, "Touching a woman inappropriately on the first date will get you further with her than not touching her at all. Don't let a woman's faux indignation at your boldness sway you; they secretly love it when a man aggressively pursues what he wants and makes his sexual intentions known."[54] Ken "TofuTofu" Hoinsky writes in part 7 of "Above the Game," a popular post on the subreddit r/seduction ("Seddit," a community that currently has over 250,000 subscribers), that

> all the greatest seducers in history could not keep their hands off of women. They aggressively escalated physically with every woman they were flirting with. They began touching them immediately, kept great body language and eye contact, and were shameless in their physicality. Even when a girl rejects your advances, she KNOWS that you desire her. That's hot. It arouses her physically and psychologically.[55]

When Hoinsky attempted to turn the post into a book by funding it on the popular crowdfunding site Kickstarter, he received a huge amount of backlash from the feminist community; as a result of the controversy, Kickstarter banned all future seduction manuals from their site. Hoinsky was shocked at this response, even though later sections in the guide contained such self-evidently problematic sentiments such as *"Don't ask for permission. Be dominant. Force her to rebuff your advances"* and *"Don't ask for permission, GRAB HER HAND, and put it right on your dick"* (original emphases).[56]

In kissing, as well, both Ovid's *praeceptor* and modern pickup artists argue that it is easier to ask for forgiveness than permission. Ovid's narrator instructs his reader,

> So she doesn't give them. Take them without her giving!
> She'll fight back, perhaps, at first, and call you "Monster!"
> Fighting back, she still wants herself to be conquered.
> Only take care that kisses snatched the wrong way don't
> injure
> Her tender lips, that she can't complain they were hard.
> (*Ars* 1.664–668)

Kissing is also a major preoccupation in the game community, because it constitutes what is known as a "phase shift" from flirting to actual sexual contact. Hoinsky advises his readers, "One common mistake men make is waiting until the end of the night to go for the kiss. . . . It's much more effective to sneak a kiss in during the date, at the first opportune moment. It diffuses the awkwardness of sexual tension. Grab her and kiss her. Sneak it in when she least expects it."[57]

All pickup artists argue *against* asking a woman whether she wants to be kissed. The man's desire to kiss the woman is deemed more important than her desire (or lack thereof) to be kissed. In *The Game*, Strauss includes a transcript of an online chatroom in which he asks some fellow pickup artists for advice because "transitioning to the kiss is a big hurdle for me."[58] Von Markovik's advice for overcoming this hurdle is, "I don't just say 'I don't care what she thinks.' I actually don't care what she thinks."[59]

Unsurprisingly, the game community was completely unfazed by the 2005 Access Hollywood recordings of Donald Trump telling Billy Bush, "I'm automatically attracted to beautiful women—I just start kissing them, it's like a magnet. Just kiss. I don't even wait."[60] To a pickup artist, this behavior is natural, even aspirational. On

Chateau Heartiste, Weidmann often calls this mindset "zero fucks given" (ZFG). The idea is that, when you stop caring about whether a woman actually *wants* you to kiss her, initiating contact becomes much less intimidating.

For shy, beginner pickup artists, instructors usually recommend starting with "night game," because under the cover of darkness, women are more likely to be intoxicated and therefore have more permeable boundaries. Similarly, Ovid's *praeceptor* advises his students to attend dinner parties where their targets will be drinking (*Ars* 1.229–244). However, they should not get drunk themselves—a suggestion that anticipates the typical pickup artist line that not consuming alcohol saves money, prevents accidentally sleeping with an unattractive woman through the phenomenon sometimes known as "beer goggles," and keeps your game sharp (*Ars* 1.589–590).[61] Drinking, according to many members of the seduction community, may make you feel confident, but will make your game technique sloppy. Ovid's *praeceptor* agrees, and adds the suggestion that would-be seducers should *appear* to be drinking, because if they then seem too sexually aggressive they can blame their behavior on the alcohol (*Ars* 1.597–600).

The advice given to women about alcohol is completely different, and highlights how even the third book of the *Ars Amatoria*—ostensibly a book targeted at a female audience with advice for how to seduce and attract men—in fact focuses on loosening female inhibitions. Ovid's text advises women that, if they get drunk, then they deserve to be raped:

> A woman lying in a puddle of Lyaéus?[62] Disgraceful.
>   She's worthy to suffer any sex whatsoever.
> Nor is it safe to lie there asleep when dessert's been
>   served;
>   many shameful things often happen through sleep.
>   (*Ars* 3.765–768)

Both male and female alcohol consumption excuse sexual assault. If a woman does not stay completely sober and keep her distance from drunk men, she becomes fair game.

Nor is this the only example in *Ars* 3 with assumptions about female behavior that mirror those held by most pickup artists. More than a quarter of the book is spent advising women how to dress and style their hair in a way that will be most attractive to men. (In contrast, only a brief section of *Ars* 1 addresses male appearance, and the advice given is that men should not put too much effort into it.) Ovid's *praeceptor* emphasizes that beauty is a woman's trump card. Just as a rich man has no need of *ars*, neither does the naturally beautiful woman (*Ars* 3.257–258). This advice anticipates the pickup artist truism that while men need extensive game techniques to attract women, women only need to look beautiful.[63] *Ars* 3 also begins with an encouragement to women to have as much sex as possible, because they have nothing to lose—after all, women do not have a limited number of sexual encounters to ration out over the course of their lifetimes (*Ars* 3.89–94). This exhortation suggests that Roman women engaged in something like what pickup artists call "anti-slut defense" (ASD), a relic of a woman's social programming that causes her to fear that having many sexual partners will lower her social status. Considering that many manosphere sites recommend against attempting to form a long-term relationship (LTR) with a woman who has had more than one or two sexual partners, that fear is entirely justified.

Even the methodological differences between the *Ars Amatoria* and modern game texts can reveal a similar ethos. Ovid's *praeceptor* often gives advice that Valizadeh refers to in his review of the *Ars* as "beta game" and von Markovik would certainly characterize as "demonstrations of low value."[64] He advises the reader to send eloquent love notes, to cry in front of his target, to always be by her side (1.437–442, 455–562; 1.659–663; 1.487–504). All of these actions would be perceived by modern pickup artists as confirming the woman's higher value and the man's corresponding lower value. Most of all, Ovid repeatedly tells the reader to compliment his target effusively, although you should never go so far that you might get caught in a lie (1.621–630, 2.295–314). On the other hand, he thinks that euphemisms are ideal—if she is so emaciated that she is barely alive, he says to tell her she is "slender" (2.641–662, 2.660; *Sit gracilis, macie quae male viva sua est*). He also tells his reader to be persistent; if you do not give up, with enough time you could even win over Penelope herself (1.477; *Penelopen ipsam, persta modo, tempore vinces*). In contrast, most pickup artists today advise that if a pickup is going badly, it is easier to leave and try again on a different woman than to attempt salvaging it.[65]

Ironically, the *Remedia Amoris*—Ovid's sequel to the *Ars*, a guide to falling *out* of love—has at least as much in common with modern seduction ideology as the *Ars* itself. In a direct reversal of the advice from the *Ars* for how to compliment your target's flaws, the *Remedia* tells how to turn her most attractive attributes into faults: "in slenderness there's the charge of being lean" (*Remedia* 328). This section of the *Remedia* could serve as a textbook on how to deliver a neg, and it resembles closely the tenth of Weidmann's "Sixteen Commandments of Poon":

> The man who trains his mind to subdue the reward centers of his brain when reflecting upon a beautiful female face will magically transform his interactions with women. . . . It will help you acquire the right frame of mind to stop using the words *hot, cute, gorgeous,* or *beautiful* to describe girls who turn you on. Instead, say to yourself "she's interesting" or "she might be worth getting to know." Never compliment a girl on her looks, especially not a girl you aren't fucking. Turn off that part of your brain that wants to put them on pedestals.[66]

Ovid also tells his reader that aloofness can itself be a powerful seduction strategy in the *Remedia*:

> She'll stop putting on airs when she sees you're growing
>   lukewarm
> (here's yet another service you'll get from my *Art!*)
> . . . .
> She mustn't be too pleased with herself, or scornful of
>   you:
> buck up your spirit, so she'll yield to your spirit.
>   (511–512, 517–518)[67]

These pieces of advice resemble two of Weidmann's other suggestions: commandment II, "Make her jealous," and commandment
VII, "Always keep two in the kitty."[68]

The *Remedia* has a long history of being used, especially in the
Middle Ages, as a text to teach young men how to stay chaste and
not be overcome by passion, so it is ironic that it so resembles the
techniques men now use to have as much casual sex as possible.[69]
Yet game principles today ultimately rest on the advice von Markovik gave Strauss for how to stop being afraid to initiate a kiss:
stop caring what she thinks. If you stop caring, then the sting of
failure goes away, and in all likelihood women will find your blasé
attitude attractive and try even harder to win your approval.

In this attitude more than any other, Ovid's *praeceptor* is someone
pickup artists might wish to emulate. He has so deeply internalized
the idea that women are inferior that he does not need to instruct
the reader not to care what she thinks. Even when he advises his
student to shower compliments on his target, the *praeceptor* warns
that such effusions will bring shame (*Ars* 1.621). He advises, "Don't
think it's shameful for you (shameful, yes—but charming) / to
hold up her mirror with your well-bred hand" (*Ars* 2.215–216). After all,
the student is an elite Roman citizen male, and from his perspec-

tive the target is just a *meretrix*—an expensive whore, to be sure, but a whore nevertheless. Why should he demean himself to her? Why should his well-bred (*ingenua*, signifying that he is free-born and not from a slave background) hand hold up a mirror so a whore can preen herself? But the *praeceptor amoris* encourages his reader not to worry. The social differences and hierarchies will not be permanently overturned if you pretend she is superior to you for long enough to convince her to sleep with you for free.

Although only the "Darth Vaders" of the seduction world have successfully assimilated their thinking about female personhood to Ovid's, most pickup-artist instructions aim at producing an Ovidian dynamic between the aspiring seducer and his target. Ovid's narrator advises giving compliments because he is completely secure in his superiority. Today's writers such as Valizadeh and Weidmann advise *against* giving compliments because they know that many of their readers feel an initial impulse to simultaneously objectify women and verbally reward them. Superficially, these may be mutually exclusive pieces of advice, but they operate under the same principle.

Put in this light, the game community's adoption of Ovid as the father of seduction looks more sinister—even before Ovid's *praeceptor* explicitly advocates sexual assault.

A Dangerous Game

Many members of the manosphere feel disenfranchised by and disgusted with the amount of societal power women allegedly hold. Pickup artists are obsessed with the declining quality of women in the West in particular; in "The Decimation of Western Women Is Complete," Valizadeh writes,

> It's amazing that in just three generations, women have gone from being potential wives and mothers to nothing

more than fuck toys. Men used to meet traditionally minded virgins, but are now stuck with a seemingly unlimited pool of mediocre sluts who have been fucked in the ass by multiple men. This is complete and utter decimation of the female human. Men can no longer gain any meaning or value from a woman beyond sex, even if he [sic] is truly capable of being the world's number one dad, and rest assured that the degeneracy that has so swallowed America whole will spread throughout the world from the trojan [sic] horse technology out of Silicon Valley.[70]

In the next chapter, I will discuss what the Red Pill community considers an ideal state of sexual politics, or what Valizadeh is thinking of when he talks about "potential wives and mothers." But the pickup artist does not live in an ideal world; his mission is to enjoy as much sex as possible in this era of moral depravity.

Pickup artists do not think that using game tactics is manipulative. Instead, they see them as a necessary weapon in an escalating arms race between the sexes, in which women hold several distinct natural advantages. From a pseudo–evolutionary psychology standpoint, they consider it in a man's best interest to have sex with as many women as possible, whereas it is in a woman's best interest to find and hold onto the highest-value man to provide for her and her offspring. Men therefore naturally want sex more than women do, which means that women control the resource supply. On top of that natural biological advantage, society also trains women from a young age to make themselves as attractive to men as possible through the use of makeup that enhances their physical attractiveness.[71] Men *need* game to catch up. Learning game, therefore, is a way of subverting the "traditional" hierarchies of sexual politics, which they claim favor women. For these men, seduction is very much a subversive political act, in addition to a lifestyle and a means to a gratifying end.

It is in this self-conscious politicization of sex that the pickup artists are most similar to Ovid. The sex acts that Ovid gives advice on how to obtain were very nearly illegal when the *Ars* was published. In a series of laws perhaps aimed at encouraging higher fertility rates among the upper social classes in Rome, Augustus instituted financial benefits for multiparous couples and financial disincentives for long-term bachelor- and bachelorette-hood (the *lex Iulia de maritandis ordinibus*, 18 BCE), and he made adultery punishable by banishment (the *lex Iulia de adulteriis coercendis*, 17 BCE).[72] The Roman historian Tacitus makes Augustus's seriousness about this legislation clear in a story about how the emperor invoked the latter law against his own daughter and granddaughter (Tacitus, *Annals* 3.24). As classicist Alessandro Barchiesi puts it, "Anyone who sets out with the idea that the *Ars amatoria* is a frivolous text, because it describes adulterous love affairs and sensual pleasures, runs the risk of not realizing the fundamental importance that the areas of morals and private life have for the new regime."[73]

Ovid conspicuously denies breaking any laws in the *Ars*, but by writing a mock seduction manual in an age of restrictive sexual morality, he made himself into a true model for turning casual sex into political resistance. Regardless of how seriously Ovid intended that resistance, Augustus punished him for it severely. Ovid ended his life in Tomis, a city in modern Romania, where he was exiled in 8 CE because of, in his words, "a poem and a mistake." The poem in question is, of course, the *Ars Amatoria*: "songs made Caesar put a black mark on me and my morals— / because of my *Art*, published so long ago" (*Tristia* 2.207, 2.7–8).

Although Ovid's late poetry is filled with complaints about his exile in a barbarian land and requests to be allowed to come home, he is surprisingly sparing with the details of his exile. It also does not seem to have occurred to him to stop writing poetry, even though poetry was supposedly the reason for his exile; instead, in *Tristia* 2, Ovid presents a long letter to Augustus that functions as a programmatic model for how readers "ought" to read his poems.[74] Ovid makes repeated claims throughout the poem that the *Ars* has been taken too literally, and that if Augustus really read it instead of relying on reports of it, he would see that the poem didn't encourage illegal behavior (*Tristia* 2.239–252).

It is debatable whether Ovid offers advice that truly contradicts the *leges Iuliae* in the *Ars*, although he certainly fails to follow the spirit of the law, if not the letter. But nothing in the text mocks Augustus's moral legislation as blatantly as Ovid's fellow elegiac poet Propertius did years earlier when he openly rejoiced at the failure of a proposed bachelor tax:

> Cynthia, it certainly delights you that the law did not
> pass,
> that decree which made us both weep for so long,
> Since it could have divided us: although not even Jupiter
> himself
> Could divide two devoted lovers from each other.
> "But Caesar is powerful." Powerful he is, in military
> strength—
> Conquered nations have no power over love.
> I'd rather have my neck separated from my shoulders
> Than betray you by getting married.
> (Propertius 2.7.1–8)

Ovid makes no such pronouncement, and strictly advises *against* fidelity to any single woman, as Weidmann does with his "poon commandment" to "keep two in the kitty." But Ovid's subversion may be greater than Propertius's, since he gives advice for sexual activity that could be perceived as directly contravening an existing law rather than celebrating the failure of a defunct law. Although Ovid's advice is ostensibly aimed at the seduction of *meretrices*, he could also be advocating and facilitating illegal acts, such as the seduction of a married woman. Furthermore, in the decades between Propertius's poem and Ovid's, Augustus's control over the population and political apparatus had only grown stronger.

Even worse, Ovid seems throughout the *Ars* to slyly needle Augustus and perhaps also his wife Livia.[75] Several of the monuments suggested by Ovid's *praeceptor* in *Ars* 1 as prime day-game spots for picking up women were built by Augustus to celebrate the imperial family; Ovid also suggests that triumphs and mock naval battles put on in honor of Augustus's military victories are opportunities to run into desirable women. It is easy to see why Augustus might have felt about the *Ars* the way we do about internet guides for how to build pipe bombs: the publication of the document itself may not be illegal, strictly speaking, but it certainly facilitates illegal activity.

There is no evidence outside of Ovid's own poetry for his exile to Tomis, so some scholars have doubted whether it actually happened. If not, it is a fascinating fiction, positioning seduction and the teaching of seduction as something so dangerous and threatening to the political order that it could potentially get you sent to the other side of the world. In his exile poetry, Ovid ultimately represents himself as a new kind of "average frustrated chump," the man hopelessly in love with his fatherland:

> Some sorrows cannot be cured by skill (*arte*)—
>> At least, not until after a long passage of time.
> When your advice has strengthened my faltering spirit
>> And I've girded my heart with your armor,
> Then my love for my Rome comes back, stronger than all
>> reason
>> And it undoes any progress your words had made.
> Call it whatever you like—duty, effeminate (*muliebre*)
>> emotion.
>> I confess that my miserable heart is weak and soft
>> (*molle*). (*Ex Ponto* 1.3.25–32)[76]

Ovid is unable to use his own advice from the *Remedia Amoris* to lessen that love's power over him.

It is ironic that, while Ovid was sent against his will to the Black Sea, many pickup artists now actively choose to go to Eastern Europe in pursuit of a higher caliber of woman than can be found in the United States. Valizadeh even wrote a book called *Poosy Paradise* about his own sexual experiences in Romania in 2014. But that renunciation of American women, even to the point of expatriation, is itself a political statement in its vocal rejection of contemporary US sexual norms.[77] So is the white nationalism of Weidmann, who fuses sexism and racism when he writes about "the intuitive understanding existing in all people (even nonWhites) that White woman pussy is the Moloko Bush of earthly pussy" and asserts that a white woman's "alabaster skin is a bucket of boner bait no other race of women can simulate."[78]

A majority of members of the pickup artist community are white, mirroring the demographics of the Red Pill more generally.[79] That whiteness is the default is obvious from how pickup artists of color center their race as a primary factor in their choice of which game techniques to use. Mumia "Obsidian" Ali, one of the three bloggers framed by Galbi as a latter-day Ovid, has fashioned a position for himself within the manosphere as a critic of black feminism and a guide to how seduction works differently within the black community.[80] Donovan Sharpe, a writer for *Return of Kings*, writes about strategies black men should use if they want to sleep with white women.[81] Valizadeh, the publisher of the site, is half Armenian, half Iranian—a racial makeup that generates tension in his alliance with the Alt-Right community, who frequently pelt him with racist slurs. However, he frequently refers to his Christian faith, presumably to subtly remind the intensely Islamophobic Red Pill community that he is not Muslim.

Microaggressions and Microeconomics

One of the most frequent criticisms leveled against the seduction community by the rest of the manosphere—especially the Men Going Their Own Way (MGTOW)—is that pickup artists are completely dependent on female approval. The precise phrases often used are that they are "slaves to pussy" and "put pussy on a pedestal." By focusing so much energy on making themselves appealing and attractive to women, these critics claim, the seduction community is complicit in a gynocentric society that inherently values women more than men. They may *claim* to subvert sexual politics, but they really confirm the status quo.

In early 2016, there was a brief uproar in the Medieval Studies academic community when the online writings of Allen J. Frantzen, a respected professor of Medieval Studies at Loyola University Chicago who retired in 2014, went viral. Frantzen explicitly invoked the concept of the red pill. However, he also seemed somewhat disenchanted with the manosphere at large and complained in now-deleted blog posts, "The manosphere isn't about men. It's about women and what they want. . . . Bloggers talk about how to get women's attention and admiration and get them into bed. . . . These writers are not thinking about who men are or what men need."[82]

Frantzen's insight here is a trenchant one, even though it leads him to the questionable conclusion that we live in a "feminist fog": "the sour mix of victimization and privilege that makes up modern feminism and that feminists use to intimidate and exploit men."[83] But he is correct that much of the discourse in the manosphere, a community that is supposedly aimed at men, is actually about how to successfully navigate various interactions with women.

The sheer volume of words published on the internet about seduction techniques—and the time and energy dedicated by the seduction community *en masse* to analyzing how to obtain access to female orifices—suggests that they do assign a high value to the act of having sex with a woman. But there is a crucial difference, ignored by Men Going Their Own Way, between valuing *women* and valuing *sex*, or rather sexual encounters with women. The fact that women constitute such a large part of game discourse should not be taken as a sign that pickup artists highly value women. On the contrary, the process of obtaining sex has been so heavily theorized by the seduction community that the subjectivity of human women has been effectively removed from the equation. These are men talking to other men and competing with other men about their success at obtaining a valued commodity.

Close examination of the dynamics of the seduction community reveals that picking up women is in many ways less important than the bonds and rivalries between the community members. To an outsider, it may look like a group of men talking about women, but often the women become little more than a means to the end of establishing authority and social capital among a group of male peers. Although the Men Going Their Own Way claim that pickup artists need affirmation from women, the female gaze is not a primary concern in modern seduction ideology. Female appearance is paramount, but the pickup artist is encouraged to improve his own appearance to make him more confident, not just to make him more attractive to women. The gaze that matters most is his own, followed by the gazes of others like him. Strauss writes of his rise from student to master of seduction that "before I joined the community, I had been afraid of failing in front of women. Now I was afraid of failing in front of men."[84]

This homosocial element is even implied in some recommended seduction techniques. Von Markovik's standard protocol for "opening a mixed set"—that is, insinuating oneself into a group of people that includes both men and women for the purpose of seducing one of the women—is that you should start by trying to impress and win over the men.[85] This maxim is especially true if the man is a threat, which is to say that he is an alpha male of the group (AMOG) rather than a hapless beta male. These tactics imply a set of assumptions about how social interactions between men and women work. Men are always in control, even if women are the ones with the power to grant sexual access.

Ovid's poetry is likewise an expression of elite Roman masculinity. In a sense, the *Ars* and *Remedia* are about how young men in Rome should, or perhaps should *not*, have sex. Even the third book of the *Ars Amatoria*, which claims to offer advice to women on how to seduce men, often seems to be engineered for the male audience's benefit. They are books for a primarily male audience—and one specific, very important male reader: Augustus. They are the product of a narrator and a poet constantly thinking on multiple levels about how to outmaneuver other men—how to make an ally of and then replace the *meretrix's vir* (her primary customer, or perhaps her husband), and how to establish the place of his own work at the top of the genre of love poetry written by male elegiac poets.

Perhaps unsurprisingly, anything that reinforces homosocial bonds between men opens itself to suspicions of homoeroticism. Pickup artists know very well that they pay more attention to grooming, self-care, and accessorizing than most of the heterosexual male population, and they are insecure about the ramifications of their atypical performance of masculinity. This insecurity often gives way to homophobia. Accusations of homosexuality are commonly used as insults on game message boards, and near the end of *The Game*, Mystery says to Strauss, "You are the most important man in my life. . . . Try not to queer that up, ok?"[86] And it is hard not to read a gay slur into Ross Jeffries repeatedly calling his rival and former student David DeAngelo "David DeAnushole."

Imagining modern sexuality in the United States as a gay / straight binary is more than simply reductive; it is inaccurate. But, for the most part, that is how the men of the manosphere think. A nuanced view of sexuality and gender expression is anathema to them: the only accepted modes of self-presentation are "heterosexual masculine man" and "heterosexual feminine woman." Divergence from either of these norms, however slight, constitutes perversion. The Red Pill community disdains unfeminine women, and when it comes to men, all deviations—being gay, or trans, or beta, or a stay-at-home father—are perceived as a shift toward increasing feminization (that is, being a "mangina").

Roman elegists, like today's pickup artists, perform their masculinity in unexpected and unconventional ways. Sexuality in ancient Greece and Rome, inasmuch as "sexuality" can be said to have existed, was as complex and multifaceted as it is today.[87] Elegists in Rome such as Ovid self-consciously adopted submissive attitudes toward the women to whom they address their poetry. This stance of being a slave to love (*servitium amoris*, as opposed to engaging in love as a battle, *militia amoris*) opened the elegist to the charge of being *mollis*: soft, or perhaps even effete.[88] An excess of desire for and time spent with one woman had the effect of making a man seem less masculine to his peers.

For the most part, the poets embraced this complicated and unstable construct of elegiac masculinity, but never to the point where their manhood would actually come into question. The most famous rebuttal of this sentiment is not from Ovid, but from one of his predecessors, Gaius Valerius Catullus, who lived in the middle of the first century BCE. Catullus is most famous for his poems celebrating and mourning his lover Lesbia's pet sparrow, and also for his famously romantic "kissing poems."[89] One of these reads

Give me a thousand kisses, a hundred more,
another thousand, and another hundred,

and, when we've counted up the many thousands,
confuse them so as not to know them all,
so that no enemy may cast an evil eye,
by knowing that there were so many kisses.
     (Catullus 5.7–13)

A few poems later, Catullus reacts with an aggressive expression of his penetrative sexuality to a criticism he claims to have received for those kissing poems:

You, who read all these thousand kisses,
you think I'm less of a man (*male me marem*)?
I'll fuck you in the ass, and I'll stick it in your mouth.
     (Catullus 16.12–14)

That "less of a man" perfectly expresses the fear of the poetic lover and the PUA that his performance of masculinity is insufficient.[90]

This fear of seeming effeminate or homosexual, shared by modern pickup artists and Roman elegists, seems to rest on the subtle awareness that teaching other men how to seduce women to some extent requires a seduction *of the student* by the teacher. The aim of the person who writes a game guide is not in fact to seduce women. He has already accomplished that goal, although he almost certainly intends to continue doing so in the future. But the guide itself is unlikely to aid in those aims; in fact, women may be repelled if they learn about a man's professional involvement with pickup artistry.

Instead, the guide itself exists for the approval and purchase of the other men who will be its readers. The same skills that they teach to help with a successful pickup—telling entertaining stories, displaying value—are used in the text on the reader, both to convince him of the writer's success with women and to convince him that the writer is different and more appealing than other aspiring pickup teachers. Neil Strauss's mantra is "The pickup artist must be the exception to the rule," and throughout *The Game* he subtly establishes and reiterates his own difference from the rest of the seduction community. Unlike the rest, he is articulate and literary. He declares early on, "I am a deep man—I reread James Joyce's *Ulysses* every three years for fun."[91]

To his students, the modern *praeceptor amoris* almost becomes like the *meretrix* whom the *Ars* aims to help the reader become more skillful with. The student has to be drawn in by the *praeceptor*'s accessible, relatable manner, then captured by his intelligence and skill. The student begins to feel that, with the *praeceptor* by his side, the world has much more potential. And, of course, the student is willing to pay for the chance to fulfill that potential.

An entire economy has sprung up around men telling other men how to be successful with women. There are books, some with exorbitant prices, including Krauser's *Daygame Mastery*. Other pickup artists run websites dedicated to articles about game and collect income from advertising revenue. The most entrepreneurial in the community become professional dating "coaches." Even an unpaid job as a moderator of a forum comes with social capital, and social capital can often be parlayed later on into more tangible benefits, such as book sales.

For the Roman poet, success and popularity could sometimes lead to gaining the support of a patron who would then provide financial security. Horace and Vergil both had the same patron, Maecenas, and Horace is effusive in his displays of gratitude after Maecenas gives him a small country villa. In a sense, the position of the poet then becomes not that dissimilar to that of the *meretrix* herself.[92] Classicist Trevor Fear states this paradox well: "Elegy, in fact, can be perceived as operating on two levels of literary prostitution: the elegiac narrator attempts (generally to no effect) to prostitute his poems for sex within the text, and, at the same time, the external narrator is using the presentation of a venal woman to prostitute his poems outside the text."[93]

Ovid's own economic model is less transparent, although he is never coy about his desire for social capital:

> Celebrate me as a prophet, men, to me sing praises:
> let my name be chanted throughout the whole world!
> I've given you arms! Vulcan had given arms to Achilles:
> conquer, as he conquered, with the presents given.
> But whoever has overcome an Amazon with my sword,
> should inscribe on the spoils, "NASO WAS MY
> TEACHER"! (*Ars* 2.739–744)

At the end of book 3 of the *Ars*, Ovid makes almost the same request of his female audience: "As once the young men did, so now the girls in my crowd / should inscribe on the spoils, 'NASO WAS MY TEACHER'!" (*Ars* 3.811–812). One might wonder if Ovid is more interested in women as objects of seduction or as readers and students (and buyers) of his poems.[94] He even explicitly advises women that, if they wish to appear educated, they also should familiarize themselves with the *Amores* and *Heroides* (*Ars* 3.339–346).

Being able to claim a female audience, for Ovid, must have been primarily a matter of social capital and posture. The *Remedia Amoris* fulfills a similar function—it allows him to boast that he is a teacher not only of love, but also its opposite. The real audience of the *Ars Amatoria* was elite Roman men.

When No Means Yes

In *The Game*, Strauss gives extensive backstories for almost every pickup artist he mentions. There are detailed, almost comically lyrical descriptions of their appearances—he describes one of the men he "sarges" with (that is, goes out with looking for women to pick up) "Grimble," in this manner: "He had the complexion of barley tea, though he was actually German. In fact, he claimed to be a descendent of Otto von Bismarck. He wore a brown leather jacket over a silver floral-print shirt. . . . He reminded me of a mongoose."[95] Strauss tells the reader their histories as "average frustrated chumps" and what motivated them to get involved with the seduction community, then catalogues their metamorphoses into successful pickup artists. In contrast, he is extremely sparing with detail about any of his conquests, with the exception of Lisa, the woman he falls for at the end of the book who inspires him to leave the community. Aside from her, most women are identified by their first names, their occupation, and their hair color.

This lack of interest in the lives of the women he sleeps with is typical of the writing in the game community. Considering the amount of doxxing (revealing someone's personal information, such as home address and social security number) and revenge porn (sharing naked pictures or video of a former sexual partner without the partner's explicit consent) that takes place in the manosphere, it strains credulity that they are trying to be gentlemen and protect the anonymity of the women they have sexual intercourse with. A more plausible explanation is that they do not care much what happens between a woman's ears, as long as they can get between her legs.

Strauss is aware of his lack of interest in the details of his conquests' backstories: "Even when I was having a deep conversation, learning about a woman's dreams and point of view, in my mind I was just ticking off a box in my routine marked rapport. In bonding with men, I was developing an unhealthy attitude toward the opposite sex. And the most troubling thing about this new mindset was that it seemed to be making me more successful with women."[96] This is not a sign of self-awareness; Strauss is not truly worried about the *women* whose substance and humanity he has effectively stripped away. Instead, he is obsessed with the potential damage he might be doing to himself.

This self-centeredness pervades his book. He says of one of the last women in the book with whom he has sex, "She was all holes: ears to listen to me, a mouth to talk at me, and a vagina to squeeze orgasms out of me."[97] His self-absorption continues in his follow-up book *The Truth*, in which he renounces his womanizing ways. After cheating on the woman he loves, he sets out on a journey of sex-addiction therapy, polyamory, swinging, and eventually a sexually adventurous but emotionally empty open relationship before finally deciding that the only thing he really wants is a monogamous relationship with the woman he betrayed.[98] Tucker Max has a similar narrative arc. His 2015 book *Mate: Become the Man Women Want* clearly positions itself as a follow-up and reversal of his 2006 book *I Hope They Serve Beer in Hell*, with the cover image from the earlier book—Max next to a woman whose face has been cut out and replaced with the words "your face here"—flipped to a picture of Max's wife standing next to a man with his face similarly obscured.

That Strauss and Max found themselves dissatisfied with their lifestyles is hardly a surprise. When discussing women, members of the community rarely mention first names, professions, or personality traits. They identify the target as HB (*hot babe/bitch* or possibly *honey bunny*), a number between 1 and 10 designating attractiveness, and possibly a physical attribute that distinguishes her. Even when a member of the community advocates for pursuing 6s or 7s over 10s, as Ali does—although he sometimes uses the term *dimes*, slang in the black community for a 10—the reasoning is not that one ought to look deeper beneath the surface. Instead, Ali argues in a post on his blog *The Obsidian Files* that by lowering your standards, you will get rejected less and laid more, which will improve your confidence and may allow you to attract more beautiful women in the future.[99] When women morph into HB8 Asian and HB7 redhead, is it any wonder that they stop seeming like human beings with the same level of personhood as men?

Roman elegy has a similar tendency to erase female subjectivity. Elegists gave their muses nicknames, theoretically to protect their identities but really to mark them in their primary role as sources of poetic inspiration. Catullus's Lesbia and Ovid's Corinna are both named after female poets (Lesbia means "the woman from Lesbos," namely Sappho); Tibullus's Delia and Propertius's Cynthia are both named for epithets of Apollo, the god of poetry. Even as these women are portrayed as holding tremendous power over the men who love them, the poetry itself serves as an instrument of male control, and the poems emphasize male erotic experience. Women are desired, praised, reviled, and longed for, but they are rarely given inner lives.[100] In *Amores*, when Ovid berates Corinna for getting an abortion, he speculates that her motive was that she did not want stretch marks on her stomach (2.14). But the reader is never given her perspective—only Ovid's anger at such a decision being made without his approval.

When seduction is the only goal, female consent becomes another barrier to be surpassed or sidestepped. In several episodes in the seduction memoir *30 Bangs*, Valizadeh is completely transparent about this mentality: "It took four hours of foreplay and at least thirty repetitions of 'No, Roosh, no' until she allowed my penis to enter her vagina. No means no—until it means yes."[101] Since the resulting sexual act is technically consensual, it would be difficult to classify it as sexual assault—but it certainly reveals that Valizadeh believes he is qualified to judge that some refusals to consent to sex should not be taken seriously.

Ovid's similarly cavalier disregard for female consent in the *Ars Amatoria* has often shocked scholars and theorists. In addition to the examples I have already mentioned of his certainty that male desire will lead to female acquiescence and his assurances that women *want* men to steal kisses from them, Ovid eventually brings this logic to its inevitable conclusion by telling the reader that women want to be forced, and even their objections are disingenuous:

> Anyone who's grabbed kisses and doesn't grab the rest too
>   deserves to forfeit even what he has been given.
> After the kisses, you almost made it to the grand prize!
>   Ah me—that wasn't modesty, it was rusticity!
> Call it force if you will, that force is pleasing to girls;
>   what they like, they often want to have given
>   "unwilling."
> Whoever's been violated by a sudden snatching of
>   Venus
>   rejoices, and wickedness is considered a service.
> But she who, when she could have been forced, departs
>   untouched?
> Though she simulates an expression of joy, she'll be
>   sad.
> Phoebe suffered force, force was used on her sister,
>   and both the snatchers were pleasing to the snatched.
>   (*Ars* 1.668–680)

When Ovid says that a woman "rejoices," the word is *gaudet*—often a euphemism for the female orgasm. The *praeceptor* is advising men that rape brings women sexual pleasure.

Some scholars believe that Ovid is insincere here, that he is mocking the reader and toying with his expectations. Roman elegy is full of rape "scripts," where the elegist talks about rape in hypothetical terms, and Ovid's advice here could be read as a kind of self-conscious commentary on this trope.[102] Ovid is, after all, a noted expert at seeming to simultaneously endorse and subvert cultural mores. But whether or not we declare that Ovid cannot be serious, one must remember that the manosphere believes Ovid's ideas here are factually accurate—that women often *want* to be raped.[103]

Valizadeh seems to be operating under the same principle as Ovid's *praeceptor* in the blog post "When No Means Yes." He writes, "While every feminist likes to repeat the phrase 'No means no,' it depends on context." He continues,

> "No" when you try to take off her jeans or shirt means . . .
> "You need to turn me on a lot more."
> "No" when you try to take off her bra means . . . "Try again in five minutes."
> "No" when you try to take off her panties means . . . "Don't give up now!"[104]

This last phenomenon is known in the seduction community as "last minute resistance" (LMR), the tendency of women to attempt to stop a sexual encounter to avoid being perceived as a slut. There is a wealth of information available online for how to "punch through" last minute resistance, because pickup artists assume that this resistance is a hurdle to surpass, not a sign that the woman is unwilling to consent to sex. Valizadeh goes on in the same post to argue that, if a woman willingly removes her clothes around a man, she should expect sex to occur because "the average man can't stop due to his innate weaknesses as an animal whose entire existence depends on him successfully mating."[105]

The pernicious aspects of the seduction community's tendency to try to legitimize itself through pseudoscience are obvious in their belief that when any degree of female arousal is combined with the natural male sexual drive, the result cannot be other than consensual sex.[106] Valizadeh is just as explicit in *30 Bangs*: "I had free reign [*sic*] to do whatever I wanted as long as that attraction was there. That's why you can sometimes stick your finger in a girl's anus and she'll politely remove your hand if she doesn't like it instead of bitching you out."[107]

Whether that amounts to sexual assault may depend on the perspective of the woman to whom it happens. Even if she feels violated afterward, which is not difficult to imagine, it would be extremely difficult to litigate. Regardless of legal definitions, Valizadeh's personal definition of what constitutes rape seems to be incredibly narrow: in addition to arguing that arousal constitutes consent to anal fingering and removal of clothes constitutes consent to sex, he has also argued that sex with even very drunk women does not constitute rape.[108]

By his own admission, Valizadeh has committed acts that legally qualify as sexual assault, including sleeping with extremely drunk women, although he maintains that "the accusation that I'm a rapist is a malicious lie."[109] He is far from the only rapist in the seduction community: in *Confessions of a Pickup Artist Chaser*, Thorn describes a "lay report" she once read on the Fast Seduction website forum that describes what is almost unquestionably a rape.[110] And, in 2015, three pickup artists living in San Diego—Alex Smith, Jonas Dick, and Jason Berlin—were charged with rape. Smith chose to go to trial and was found guilty; the other two pleaded guilty. They were users of the Real Social Dynamics pickup artist forum and specialized in what they called "train game," in which, as soon as one man finishes having sex with the woman, his friend immediately takes over, often without asking the woman's permission.[111] Smith wrote on the forum, "She will usually briefly freak out. . . . Have your buddy come in and start doing whatever on her, escalating up, then just hop off and have your buddy continue."[112] He and Jonas Dick were employed as dating coaches for a company called Efficient Pickup.[113]

If the point of game were, as its apologists claim, to teach men how to be skilled with women, then rape obviously would not be necessary. The cavalier attitudes toward sexual assault found in both Ovid and modern pickup artists show that the aim of these texts, no matter what they plead, is *not* to teach men how to become attractive to women. Rather, it is to teach them that male desire is more important than female boundaries—that what women want, think, and consent to is immaterial.

Von Markovik's original name for "The Mystery Method" was "The Venusian Arts."[114] This name is programmatic because the idea that humans are designed to survive and replicate is foundational to his worldview. The skills that help aid your survival are the martial arts, named for Mars, the Roman god of war, while skills that help aid your replication are the Venusian arts, named for Venus, the Roman goddess of love.

Ovid also thought of his work as a conflict between Mars and Venus. He begins the *Amores* by claiming that he intended to write about war, but Cupid stole one of the feet from his dactylic hexameter and turned it into elegiac couplets (*Amores* 1.1.1–4). In the *Ars*, it becomes clear that for Ovid's narrator, love itself is a battlefield: male lovers are often compared to the heroes of the Trojan War, and female lovers are compared to Amazons (*Ars* 3.1–2).[115] When, in the *Remedia Amoris*, Ovid teaches men and women how to recover from love, he imagines Cupid looking at the title of his book and assuming that Ovid is "preparing battles" against him (*Remedia* 2; *Bella mihi, video, bella parantur!*).

Following this war theme, over the course of *Ars* 3, Ovid's narrator admits to worrying that, by giving advice to women on how to seduce men, he's effectively arming the enemy: "Let's hand everything over (we've opened the gate to the enemy) / and, in our faithless betrayal, let's keep faith" (*Ars* 3.577–578). After Venus demands that he write a book of instruction for women on how to seduce men, Ovid agonizes,

What am I doing—am I crazy? Why meet the enemy with
   chest
exposed and give evidence to betray myself?
The bird doesn't show the bird-catcher the best place to
   catch it;
the doe doesn't teach the hated dogs to run.
   (*Ars* 3.667–670)

But if women are the enemy when it comes to seduction, then the first two books of the *Ars* would seem to be more damaging than the third. After all, a woman reading the first two books might learn to recognize and avoid men's seduction techniques, while a woman using the advice in the third book is clearly a willing participant in the game. Furthermore, the first two books of the *Ars* would seem to give advice to the real enemy—that is, the other men who might end up as competition for a target's attention.

The obvious response to this objection is that Ovid is not entirely serious. While it is true that Ovid almost never seems to be entirely serious, the claim that men and women are in some sense at war reveals the assumptions about gendered interaction in the text. It might even seem to point toward a similarity to the seduction community mindset in which sex is a zero-sum game, because the job of the man is to get laid while the job of the woman is to restrict sexual access.[116] Ross Jeffries has the most succinct expression of this concept: "*For guys, getting laid is a chore. For women, getting laid is a choice.*" Jeffries concludes his introduction by saying, "Enough chit chat. Let's get to battle men."[117]

For these men, sex is warfare, and women are the enemy to put under siege. Women are told that their physical appearance is their best asset, so pickup artists neg them by finding ways to subtly critique their looks. Women are warned constantly through social pressure and entertainment media not to act "crazy," so

Weidmann advises using the question "So, how normal are you?" as an opener on the social dating application Tinder.[118] Game strategies are designed to utilize women's vulnerabilities, particularly vulnerabilities that are socially conditioned—although they would of course say that women are biologically programmed to seek male approval—in order to manipulate women into sexual encounters that will afford them very little aside from a moment of validation.[119]

Valizadeh's experiences in unsuccessfully trying to seduce Danish women prove that seduction techniques work best on women who are in a vulnerable position emotionally and socially.[120] In his 2011 book *Don't Bang Denmark*, he gives what at first seems to be praise for Denmark's excellent social services: "A Danish person has no idea what it feels like to not have medical care or free access to university education. They have no fear of becoming homeless or permanently jobless. The government's soothing hand will catch everyone as they fall." He continues, "America is great if you have money, but Denmark is great for everyone."[121] Everyone, that is, except for sex tourists who have come to the country exclusively for the purpose of sleeping with Danish women.

Danish women, as it turns out, are immune to standard seduction tactics, because they do not need to rely on men for any kind of support:

> For hundreds of thousands of years, women have sought to marry powerful men with strong financial means in order to live a comfortable life (or to merely survive), but in Denmark this is not at all necessary. Danish women don't need to find a man, because the government will take care of her and her cats, whether she is successful at dating or not. Her quality of life won't be negatively affected if she happens to remain single until death, whereupon her cats will inherit her possessions according to Danish law.[122]

The result of this financial equality that derails Valizadeh's pseudo-evolutionary psychology is that Danish women are "less willing to change their behavior by adopting a pleasing figure or style that's more likely to attract men."[123] Furthermore, Denmark is "a highly feminist country. It's a place where women think they're equal or superior to men, eager to castrate them for displays of alpha masculinity."[124] Valizadeh is extremely dissatisfied with the appearance of Danish women, noting that the most attractive Danish women capitalize on their beauty by becoming models or prostitutes. As for the others, "What they do have are pussies and opinions you really don't care about hearing. That's it. Denmark takes top prize for having the most unfeminine and androgynous robotic women I've met in the world."[125] His attempts at propositioning Danish women are so unsuccessful that he decides to seek gratification instead by viciously insulting them: "I had so much resentment toward Danish women that I tried to destroy as many of them as I could in order to make the world a better place."[126]

Denmark's socialist services render its women unattractive and unattainable to sex tourists. Danish women, according to Valizadeh, are physically and temperamentally similar to women in the United States, an assessment he intends as a grave insult. However, American women can still be gamed, because they rely on men for financial support and emotional validation. The more patriarchal a society is, it seems, the better seduction tactics will work—although this observation directly contradicts pickup artist ideology, which rests on the idea that game is more necessary as societies become more socially progressive.

As Valizadeh paints an inadvertently utopian depiction of Denmark, he also reveals the disturbing reality behind other seduction guides, including Ovid's. Although seduction guides claim to level the playing field between men and women, pickup artists in fact flourish in societies that oppress women, and pickup artist guides teach men how best to exploit their myriad social advantages over women for sexual gain.

By setting out models for how men should act to seduce women, these texts also implicitly script appropriate *female* behavior. Women who react appropriately, or as expected, to seduction techniques are rewarded with the descriptor "feminine," whereas those who do not follow standard seduction scripts are unnatural and unwomanly. Such guides do not only give advice to men; they also condition their readers to expect certain kinds of female behavior and give them permission to feel violently angry when women do not conform to those behavioral models. Elliot Rodger, the twenty-two-year-old who killed six and injured fourteen in Isla Vista in May 2014, was a regular poster on the currently defunct PUAHate, an online forum where men who are unable to successfully execute seduction advice stirred up each other's resentment against pickup artists and women alike.

Seduction instructors claim to teach men how to be successful with women, but they really teach how to identify and utilize women's social weaknesses. Seduction advice also empowers aspiring pickup artists to lash out at women who reject them. Even Ovid's *Ars Amatoria* teaches a reader how to abuse a *meretrix* to the exact degree that it is in her best interest to tolerate him as a potential customer. In a society where women are supported and valued, the only solution left to the pickup artist may be sexual assault. Indeed, *Don't Bang Denmark* ends with the story of Valizadeh pressuring an extremely reluctant eighteen-year-old virgin into sleeping with him and then ruminating over why seeing her blood on his sheets arouses him so much.[127] Since leftist social progress renders game tactics ineffective, it is no wonder that pickup artists long for the lost mores of ancient Rome.

———————

In "Powerlust Moves," a 2017 article on *Chateau Heartiste*, Weidmann advises men who are in romantic relationships to approach their partner from behind, wrap their arms around her, and then tell her not to turn around:

> Now she's stuck facing forward, maybe over the kitchen sink noticing tree leaves ripen in the summersun [*sic*] through the window, engulfed by my body while my patriarchy presses into her behind. I lift her dress, or unzip and yank down her pants, and explore like a White colonialist of old. All the time she is yielding to my loving molestation, her back is to me; she never locks eyes. This combination of male entitlement, commanding presence, and her sensual vulnerability is lethal to the female limbic system, dynamiting her dendritic fuses in a volcanic shower of molten gash-ash.[128]

Weidmann's evident delight in his own offensiveness aside, this fascinating passage reveals a great deal about how pickup artists think. It is obvious that Weidmann believes he has tapped into some kind of natural, biological truth: the "dendritic fuses" of her neurons subtly connect her "limbic system" to the "tree leaves ripen[ing] in the summersun." (*Dendron* is Greek for *tree*.) But throughout, he also underscores the degree to which her desire is socially conditioned: she is standing at the kitchen sink, his erection is his "patriarchy," his touch is assimilated to colonialism. Although the seduction community claims that its tactics are derived from evolutionary psychology, they in fact depend entirely on social conditioning and reflect the social context of their writers.

The attraction that Ovid's *Ars Amatoria* holds for the seduction community is easy to understand: the mere existence of a two-thousand-year-old poem that gives similar advice to what is disseminated today powerfully validates pickup-artist ideology. Ovid's status as a literary heavyweight is a substantial bonus, as is the fact that his works (such as the *Metamorphoses*, as I discussed in

Chapter 1) have already been deemed "dangerous" by so-called social justice warriors.

Unfortunately, Ovid's text is more useful to pickup artists in
theory than it is in practice. In his book review, Valizadeh calls it "a
chore to get through."[129] Weidmann assesses it as "pretty uneven in
its advice."[130] The text's utility to them extends only insofar as it
confirms their beliefs about history, gender, and sex and validates
their own "art" by giving it an ancient pedigree.

Ovid's text may have only limited value when used to validate
modern game advice, but it works beautifully to illuminate the flaws
and dangers in pickup-artist ideology. Taken alone, the *Ars Amatoria* has often been treated by scholars as an amusing literary game,
and seduction manuals are excused as common-sense tips for how
to win the game between the sexes. But when placed alongside each
other, each text appears in a different light. Ovid's casual references
to sexual assault seem far more sinister and less ironic when one
realizes that similar ideas are widespread in the seduction community today. And Ovid shows that, for certain men, the most seductive idea of all is that mastering the art of love—that is, learning how
to violate women's boundaries in a socially acceptable manner—can
function both as social commentary and as political resistance.

4

HOW TO SAVE WESTERN CIVILIZATION

In a 2017 article on *Return of Kings*, "The Problem of False Rape Accusations Is Not Going Away," the writer Christopher Leonid claimed that "having an accusation or two to your name is gradually becoming a standard rite of passage for any man of worth." Articles on the topic of false allegations are extremely common and popular on Red Pill fora, but this article was framed in an atypically literary, classical fashion: its subtitle was "The Death of Hippolytus," and the hero image for the article was the 1860 painting of the same title by the Dutch painter Sir Lawrence Alma-Tadema.[1]

The myth of Hippolytus and his stepmother Phaedra has been told and retold for millennia. The precise details of the story vary from telling to telling, but in every version, Phaedra, the wife of the Athenian king Theseus, falls in love with her stepson Hippolytus. Hippolytus is an ostentatiously chaste youth who prefers hunting to women, and when he rejects Phaedra, she lies to her husband and claims that Hippolytus raped her. Theseus then calls down the curse of his father, the sea god Poseidon, on Hippolytus. Hippolytus dies in a gruesome chariot accident and Phaedra commits suicide.

The rhetoric of juxtaposing Alma-Tadema's painting with Leonid's article will be obvious to those who have read this far in this book. The image implies that false accusations are an ancient problem that has persisted to the present day, because the urge to make punitive false accusations is intrinsic to female "nature" (a recurring Red Pill obsession) and always has been. As the Red Pill truism goes, "all women are like that" (AWALT). Additionally, celebrating the classicizing style and subject matter of an Alma-Tadema nods toward the supremacy and alleged superior quality of traditional Western European art. Regardless of whether Leonid or his editor selected the image and subtitle, their use fits seamlessly into the ethos of *Return of Kings* as a whole.

False rape allegations are an obsession within the Red Pill community across many of its factions. Daryush "Roosh V" Valizadeh, a representative of the pickup artist faction, writes about the topic often. In the 2015 post "Men Should Start Recording Sex with a Hidden Camera," he advised men to secretly film all sexual encounters to protect themselves against the possibility of a false rape allegation—even though recording someone without their consent is itself illegal in several states.[2] He also argued in a 2014 blog post, which I will revisit later in the chapter, that "All Public Rape Allegations Are False."[3]

In 2010, the men's rights activist site *A Voice for Men* published the article "Jury Duty at a Rape Trial? Acquit!" by Paul Elam. Elam, who later claimed the article was "deliberately inflammatory," wrote that "*should I be called to sit on a jury for a rape trial, I vow publicly to vote not guilty, even in the face of overwhelming evidence that the charges are true*" because "better a rapist would walk the streets than a system that merely mocks justice enslave another innocent man."[4]

A thread on MGTOW.com, an online hub for the Men Going Their Own Way sector of the manosphere, argues that "false rape is FAR worse than rape." It is a perfect example of one of the Red Pill's favorite rhetorical misdirection techniques, the false equivalence. The author of the thread argues that, in the case of the falsely accused, "*mental/emotional trauma is far longer and thus much more severe* and damaging to a person's well being"; he also believes that "not only do women NOT get harmed most of the time, they also enjoy it thoroughly. Women look for the bad boy and LOVE dominant men. They like being treated like shit and just abused by a man. (These are not stories it's a proven fact of life.)" Following this assessment that female nature inclines women to be susceptible to enjoying rape, he concludes, "Only a feminist would say that both are equal or rape is worse . . . when clearly *the severity of the consequences and time frame of false accusations are CONSIDERABLY more.*"[5]

There are so many articles, blog posts, and Reddit threads about false allegations that it is possible to identify distinct subgenres: articles that give advice on how to avoid false allegations or how to handle them if you are accused; articles that specifically address the connection between false allegations and higher education; articles that dissect public rape allegations to argue for their falsity; articles that propose appropriate punishments for false accusers.

Considering this fixation on false allegations and the Red Pill community's frequent use of classical texts to bolster their antifeminist ideology, one would expect the Red Pill to be interested in an ancient myth about a false rape allegation with disastrous consequences. But Leonid's article is a rare exception. Even though the myth of Hippolytus seems to conform perfectly to Red Pill narratives, searches of his name in the most popular Red Pill websites and subreddits return few results.[6] This lack of interest is especially surprising considering that two of the Red Pill's favorite classical writers created literary versions of the myth: Seneca, the Stoic philosopher who unsuccessfully advised the emperor Nero on how to control his emotions, and Ovid, the Augustan-era poet who was exiled to the Black Sea for writing what has been called "the oldest game book in existence." In the hands of Seneca and Ovid, Phaedra seems uniquely suited to prove that the urge to falsely accuse men of rape is a constant in female psychology throughout history.

Despite the lack of attention paid to Hippolytus and Phaedra in the Red Pill, I believe it is worth reading the ancient myth in the context of Red Pill ideology. These men are obsessed with and rather fatalistic about false allegations; as the title of Leonid's article claims, "the problem . . . is not going away."[7] But the purported inevitability of false allegations is not, as men such as Leonid and Elam claim, evidence that men are treated unfairly in our "gynocentric" society. Indeed, if the myth of Phaedra can teach us anything, it is that false allegations may occur even under the most patriarchal and restrictive of conditions. A close reading of the myth undoes the Red Pill's false and misleading narratives about the politics of gender and sexuality that produce false allegations.

Feminists usually respond to discussions about false allegations by appealing to statistics and attempting to prove that false allegations are so rare as to not be worth paying attention to.[8] Since the statistics themselves are so difficult to interpret, this tactic is rarely effective. A more successful strategy might be to challenge the Red Pill on its definitions of the underlying concepts behind their assumptions about false allegations: what does consent really mean to these men, and what does rape mean? What are the conditions under which they believe that men deserve to be punished for sexual activity? What are the sexual politics of the world they wish we lived in?

In 2015, Valizadeh published the controversial (for the time) blog post "Women Must Have Their Behavior and Decisions Controlled by Men." In that post, he posed this question: "Men, on average, make better decisions than women. If you take this to be true, which should be no harder to accept than the claim that lemons are sour, why is a woman allowed to make decisions at all without first getting approval from a man who is more rational and levelheaded than she is?" False rape allegations are, he argues, a natural consequence of granting women the unearned right to make their own decisions. "When you give a female unwavering societal trust with the full backing of the state, what does she do? *Falsely accuse a man of rape and violence out of revenge or just to have an excuse for the boyfriend who caught her cheating.*"[9]

Valizadeh's solution to the problem of false rape allegations is to give men control over their female relatives. He justifies this idea by appealing to history: "For the bulk of human history, their behavior was significantly controlled or subject to approval through mechanisms of tribe, family, church, law, or stiff cultural precepts."[10] As he suggests, and as I will return to later in the chapter, this idea is a very old one. In ancient Athens, women were under control of a male *kyrios*, typically a father or husband. In Rome, the *paterfamilias* had a great deal of control over his wife, children, and slaves.

The men of the Red Pill who write about the ancient world would have their readers believe there is a straight line from antiquity to today, a continuity of male and female behavior. As I have been arguing, however, this illusion of continuity is actually an ideologically motivated strategy to *resurrect* ancient norms in the present day. Men's rights activists, pickup artists, and the Alt-Right use the Classics to make their radical, reactionary gender politics seem not only normal and natural, but thoroughly, traditionally European.

A close examination of Phaedra's story shows how false allegations work at the nexus of cultural mythmaking about rape. The ancient myth reveals anxieties about female sexuality and female credibility that are still very much present in the United States today—anxieties that the men of the Red Pill play on and stir up. This rhetorical strategy is part of a larger project of advocating for a white society in which female consent to sex would be immaterial, and therefore a woman's claim that she did not consent—an allegation of rape—would have few consequences.

Rape Myths and Myths about Rape

Any credible discussion about false rape allegations should lead with defining the terms: what is rape, and what would make an allegation of rape false? Since the ultimate power to decide what qualifies as sexual assault falls in the hands of the government, this discussion of terminology is really a discussion about politics. To be more precise, it is a discussion about sex under patriarchy, because the feminist definition of rape and the Red Pill definition of rape are predicated on assumptions about how patriarchal our society is and how patriarchal it should ideally be.

Discussions of false allegations are always emotionally charged, because they cut close to some of our deepest discomforts about how our society deals with and thinks about rape. The grim reality is that, from a bystander's standpoint, the difference between rape and consensual sex is often not much more than one participant saying no, or even not saying yes. Rape often fails to leave a trace except on the psyche of the victim. In the absence of physical evidence, the struggle centers on whose narrative wins: she said, he said. And, as I mentioned in my first chapter, fighting against "narratives" is a central concern of the Red Pill.

One of the core beliefs of our legal system is that defendants are innocent until proven guilty. But many feminists would like to believe and support those who claim to be the victims of sexual assault. So how should we reconcile ourselves to the fact that these two stances are fundamentally irreconcilable? If we believe that alleged rapists are innocent until proven guilty, then on some level, do we also have to believe that victims might be lying until they can prove that they are telling the truth?

This apparent paradox is the natural extension of the approach to rape allegations in the Red Pill. It is also needlessly reductive, since both the accuser and the accused may be telling the truth as they see it, and there may not be an objective truth on which all parties can agree. Rather than accept this level of nuance, however, many find it easier to assume that women are natural liars. As two sociologists argue, "the term [*false allegation*] represents a rigid structure: a crude, shorthand concept in which significant attribution errors are at work . . . such attributions are guided by stereo-typical beliefs and assumptions; myths. Women 'cry wolf.' It is simply cognitively easier to work from."[11]

In other words, it is easier to fathom the idea that lying is in-herent in female nature than to grapple with the inconsistent data about false allegations and attempt to determine whether they are so rare as to be a nonexistent issue, as some feminists claim, or so frequent that men should constantly be on their guard against them, as they are represented on Red Pill websites. The best statistic avail-able about false allegations is that about eight percent of rapes are reported by individuals who are being deliberately untruthful. (A 1994 study by Eugene Kanin claiming that forty-one percent of allegations are false is frequently cited on Red Pill fora, but has been widely criticized for using extremely flawed methodology, in-cluding counting all retracted claims as false.)[12] But even that eight percent statistic, originating from FBI and Department of Justice reports from 1996 and 1997, is at best outdated.[13] Eight percent is a small number—although not so small as to suggest that false alle-gations are too rare to be worth discussing.

In "What to Do if Police Are Questioning You about a Sexual Encounter," a Red Pill writer using the pseudonym "Relampago Fu-rioso" wrote, "Women often lie to seek vengeance against men, and false rape accusations are a serious problem in our society. Women will even make up total fabrications about sexual encounters to ruin a man's name and reputation. Crime statistics prove this. If you want to get more philosophical about it, a great philosopher from a century ago called it women's innate instinct of *dissimulation*." Furioso then cites Arthur Schopenhauer, one of the manosphere's favorite misogynist philosophers: "A woman who is perfectly truthful and does not dissemble is perhaps an impossibility."[14]

Is Furioso right that "the numbers show [false allegations are] a daily occurrence"? Not entirely, since "the numbers" on false allegations are far from agreed upon.[15] Even Furioso admits that "statistics are all over the place when it comes to how many rape reports are false. Low estimates say 1–2% are false and high estimates say 90% are false."[16] Rather than striving for statistical accuracy, he invokes Schopenhauer's philosophical ideas as support for his claims—an argumentative strategy that will, by this point, be familiar to readers of this book.

Nobody can deny that false allegations exist, particularly after a few high-profile incidents in recent years—most notably the accusations made against three members of the Duke Lacrosse team in 2006 and the inconsistent story of "Jackie," the subject of "A Rape on Campus," *Rolling Stone*'s botched exposé of rape culture in college fraternities at the University of Virginia.[17] But it is impossible to tell how frequent false allegations are when we know neither how many rapes go unreported nor how many of the reported rapes really occurred.

A more fundamental concern, in fact, is that we cannot even all agree on what it means to say that a rape "really occurred."[18] Definitions of what constitutes a false allegation differ widely.[19] Is a rape allegation false only if it is made with malicious intent by someone who knows that a rape did not really occur? Or is it also false if the victim is confused about what happened and unknowingly gives misleading or incorrect information? Imagine, for example, that someone is drugged and raped at a party and is not entirely sure who did it. They report the crime to the police, giving the clearest evidence they can, and the police then arrest the wrong individual. Is that a false accusation? Some people even call *any* unverified or unverifiable rape allegation false. That usage implies that anybody who is found not guilty of rape has had a false allegation made against them, erasing the significant distance between truth and what is provable in a court of law.

The Red Pill definition of a false allegation is particularly capacious, since the community's definition of rape is so narrow. As I mentioned in the previous chapter, pickup artists believe that sometimes "no means yes," and they train themselves to expect to have to overcome so-called "last minute resistance" as the final barrier before sex can occur. In this context, it is difficult to imagine whether they would accept as "real rape" *any* kind of sexual assault that is typically classed as date rape. Unfortunately, the legal definition of rape often seems to resemble the Red Pill definition: as Judith Herman writes, "in practice the standard for what constitutes rape is set not at the level of women's experience of violation but just above the level of coercion acceptable to men."[20] The result is a definition of rape so restricted that under it, few women can claim with complete credibility that they have been raped.[21]

A post on the site SlutHate articulates the distinction between "real" and "fake" rape in especially bald terms. A man calling himself "Brother Dean Saxton" lays out what he calls "The Philosophy of Rape," saying, "My message is simple: women are property. When a woman mouths off, a man needs to beat her senseless until she shuts up. And, when harlots engage in promiscuity and other sinful behaviors, they need to be corrected with rape." He clarifies later in the thread:

> We want to teach men that although it may be easier than ever for an innocent man to be convicted of rape when a consensual partner has buyers [*sic*] remorse, it's also easier than ever for a guilty man to get away scot-free—*so long as it's done the way we advocate: actual rape-rape, as in dark-alley, ski mask, stranger rape.*[22]

The Philosophy of Rape subreddit and blog have both been shut down; even within the Red Pill community, these are extreme views. However, Saxton's definition of "rape-rape" would likely find wide acceptance in the community as the only "real" kind of rape—and the fundamental assertion underlying his philosophy, the idea that "women are property," is the foundation of a dominant strand in Red Pill (and particularly alt-right) sexual politics, as I will discuss later.

On the other side of the spectrum, some radical feminist theorists have argued (or are believed to have argued) that *all* heterosexual sex under conditions of patriarchy is rape. The men of the Red Pill enjoy mocking academic feminist theory, but they find this particular theory, closely associated with second-wave feminists Andrea Dworkin and Catharine MacKinnon, more ludicrous than any other. In reality, Dworkin and MacKinnon never made that argument, at least not in such bald terms. Dworkin actually wrote

> Intercourse occurs in a context of a power relation that is pervasive and incontrovertible. The context in which the act takes place, whatever the meaning of the act in and of itself, is one in which men have social, economic, political, and physical power over women. Some men do not have all those kinds of power over all women; but all men have some kinds of power over all women; and most men have controlling power over what they call their women—the women they fuck. The power is predetermined by gender, by being male.[23]

This position is far more nuanced than it is typically represented by the men of the Red Pill, who depict Dworkin as someone who used violent hate speech against men and was never criticized—while they are demonized for speaking their minds and not being "politically correct." One 2015 article on *A Voice for Men*, "Yes All Feminists Are like That," even calls Dworkin "the David Duke of feminism." Comparing Dworkin with the former Imperial Wizard of the Ku Klux Klan blatantly appropriates the outrage provoked by the most extreme and flagrant form of anti-black racism—a few years before David Duke became a hero to the ascendant white supremacist faction of the Red Pill community—to discredit feminist scholars working to promote gender equality.[24]

But even other feminists have characterized Dworkin and MacKinnon's arguments in similar terms: "Andrea Dworkin and Catharine MacKinnon have long argued that in a patriarchal society all heterosexual intercourse is rape because women, as a group, are not in a strong enough social position to give meaningful consent— an assault on individual female autonomy uncannily reminiscent of old arguments for why women should not have political rights."[25] Dworkin and MacKinnon's position, in this formulation, is that sex between a man and a woman is conceptually similar to sex between a slave owner and an enslaved person, or a doctor and their patient, or a teacher and their student, or a sober individual and a drunk one. The power dynamic is skewed so heavily in the direction of one party that the other cannot truly have the power to consent—or perhaps, more importantly, to *withhold* consent. Until we live in a less patriarchal society, Dworkin and MacKinnon allegedly argued, all female consent to sexual activity with men will in a sense be coerced. In MacKinnon's words, "rape has proved so difficult to define because the unquestionable starting point has been that rape is defined as distinct from intercourse, while for women it is difficult to distinguish the two under conditions of male dominance."[26]

That position, and the ideas attributed to Dworkin, are often criticized for erasing female agency and autonomy. But the idea that all heterosexual sex is rape is really something of a straw feminism, the further manipulation of ideas that are already intentionally provocative and hyperbolic. The actual question, it seems, is the degree to which we can agree that we live in a society so patriarchal that all women, regardless of class and race, are denied the possibility of autonomy and agency.[27]

Whether or not one agrees with Dworkin and MacKinnon that this description can be meaningfully applied to the United States in the present day, I think few would dispute that ancient Athens and Rome were indeed societies that severely restricted female autonomy. So was all sex in the ancient world rape? Or—as some scholars have argued—was *none* of the sex in the ancient world "rape," as we understand the word today? These questions, which may once have been of interest only to classical scholars, take on new urgency in a world where Red Pill writers are transparently attempting to resurrect the sexual mores of classical antiquity.

Rape and Marriage in the Classical World

What is rape culture? The term is often used, but its meaning is rarely agreed upon. What qualifies as rape culture in the United States today? On college campuses? In high schools with fanatical devotion to their football teams? Many men who write for Red Pill websites would strongly dispute that rape culture exists in our country at all. Instead, they point to, for example, the sexual assaults perpetrated by Muslim immigrants or "rapefugees" to prove that Islam is the true rape culture—rhetorically shifting the blame from American (and, it is implied, white) men to foreign Muslim men. Regardless, many people would probably agree on this basic definition: a rape culture exists within a social group that normalizes rape to the degree that consequences for rapists are minimal or nonexistent and punishing rapists is seen as more barbaric than rape itself.

Given that definition, it seems undeniable that rape culture thrived in both ancient Greece and Rome—although some scholars have vigorously attempted to absolve the ancients of that charge. But the desire to acquit the Greeks and Romans of complicity in rape culture is itself a revealing move, highlighting how urgently some historians feel the need to maintain critical and emotional distance from the ancient world. And no matter how one feels about the sexual politics of classical antiquity, it is plain that a woman's consent to sexual activity and to marriage was not given nearly as much weight as it is today.

The ancient Greeks did not perceive a distinct boundary between myth and history, so it is not clear that they would recognize the *Histories* by Herodotus of Halicarnassus, written around 440 BCE, as their first historical text by our definition.[28] Nevertheless, Cicero grants Herodotus the title Father of History—although he is also occasionally called Father of Lies, since the ancients were as aware as we are that representing history accurately is difficult. Significantly, Herodotus's *Histories*, the foundation on which the study of European history is built, begins with a series of rapes—or rather abductions, since in Herodotus's text the line between rape and abduction (*harpagē*) was not sharply defined.

Herodotus declares at the beginning of his *Histories* that he intends to tell the story of the Persian Wars, but in actuality his work encompasses far more material than that. It is a work not just of history, but also of anthropology, ethnography, and even travel writing. Classicist James Redfield's influential article "Herodotus the Tourist" addresses why this approach to history is so attractive and so problematic: "The tourist, in fact, travels in order to be a foreigner, which is to say, he travels in order to come home. He discovers his own culture by taking it with him to places where it is out of place, discovers its specific contours by taking it to places where it does not fit. . . . Thus cultural relativism becomes ethnocentric and serves to reinforce the tourist's own norms."[29] Herodotus's exploration of other cultures led to his defining through opposition what it really meant to be Greek, and his work had a major role in creating the concept of Greekness.[30]

When recounting the reasons for the hostility between the Greeks and Persians in book 1, chapter 1, only a few sentences after the beginning of the text, Herodotus informs his reader that the conflict began when the Phoenicians abducted Io, a woman who lived in the Greek city of Argos. The abduction of Io catalyzes a series of successive abductions, each meant to punish the other side. The Greeks abduct Europa, then Medea. Paris decides that because of these events, "he wanted to obtain a wife from Greece through abduction, because he was confident that he would not have to pay a penalty" (Herodotus 1.3.1; ἐθελῆσαί οἱ ἐκ τῆς Ἑλλάδος δι᾽ ἁρπαγῆς γενέσθαι γυναῖκα, ἐπιστάμενον πάντως ὅτι οὐ δώσει δίκας). With this reasoning, he abducts Helen of Sparta from her husband Menelaus.[31]

At this point the cycle falls apart, because the Greeks choose to respond not through another punitive abduction, but by sending an army to retrieve Helen:

> Until then it was only a matter of abducting (*harpagas*) women from each other. But from then on, according to the Persians, the Greeks were mostly to blame, because the Greeks invaded Asia before the Persians invaded Europe. They say that men who abduct women are unjust, but men who want to avenge abducted women are fools. Prudent men care little about abducted women, since it is obvious that, if they did not want to be abducted, they would not have been. The Persians say that they did not concern themselves with the women abducted from Asia—but that the Greeks, because of a Spartan woman, assembled a great army and came to Asia to destroy the power of Priam. From that point on the Persians regarded the Greeks as enemies. (Herodotus, *Histories* 1.4)

Women are not valued as people: they are mere commodities. Since the abduction of a woman is not a war crime but a property theft, waging war because of Helen is viewed as unforgivably extreme, worthy of centuries of hostility.[32]

Herodotus does not take responsibility for this viewpoint. He claims to merely report the Persian opinion on female consent to abduction, which he includes in the interest of evenhandedly displaying both sides of the argument. (Which Persians? Speaking to whom? Herodotus neglects to elaborate.) In this same spirit of fairness, he also notes that the Phoenicians tell an entirely different story about the initial abduction of Io: it was not a rape, but rather a false claim of rape to conceal Io's decision to run away with her consensual lover after she became pregnant. (The men of the Red Pill would likely prefer this version of the story, since it fits so nicely into their narrative about the female tendency to "dissemble.") Herodotus omits entirely the most familiar, although admittedly least realistic, version of Io's story, where Zeus abducted her and turned her into a cow to hide his infidelity from his wife Hera.

Is Herodotus describing what we would call a rape culture? By the definition I provided above, it would seem so, especially when he casually reports the Persian opinion that women are responsible for and co-conspirators to their own abductions ("if they did not want to be abducted, they would not have been") and Paris's blithe confidence that abducting Helen will result in no serious consequences. In fact, Herodotus is indirectly participating in two separate rape cultures. The first is the world of myth, where rape is an extremely common occurrence. The second is the world of fifth-century Greece, where the author's blasé attitude toward punishing an enemy by raping one of that enemy's countrywomen is not remarkable, but typical.

This same attitude toward rape and female subjectivity can be found in many other Greek texts. One such text, a speech by the orator Lysias, is often among the first texts students of Attic Greek read in the original language. Lysias was a speechwriter in the modern sense: he made a living writing speeches for other people. The speech in question was putatively written for a man named Euphiletus, who was on trial for murder for killing Eratosthenes, whom Euphiletus had caught more or less *in flagrante delicto* with his wife.[33] In the course of his defense, Euphiletus notes that seduction is considered a greater crime than rape. The reasoning behind this counterintuitive custom is that if you rape a man's wife, you are responsible only for damaging his property, but if you seduce her then you destroy his family by dissolving the bond of loyalty between a husband and wife and calling into question the paternity of the marriage's offspring.[34]

It seems unlikely that an Athenian wife would agree that rape was preferable to seduction, but her opinion on the matter was not a significant consideration. Nor were the opinions of the reciprocally abducted women in Herodotus. In fact, it is surprising that Herodotus does not pay to Helen of Troy's motivations, because she is the rare case of a woman in Greek myth and history whose consent or lack thereof was the subject of considerable discussion and debate. Ancient orators occasionally displayed their rhetorical skills by making hypothetical arguments about whether Helen was responsible for the Trojan War, and a key consideration in these showpiece speeches was whether her departure from Sparta with Paris was consensual or forced.[35]

But Herodotus ignores Helen's motivations entirely in favor of Paris's reasons for stealing her. In the process, Herodotus erases the subjectivity of one of the few women from Greek myth who is even occasionally afforded it. The conflict in Homer's *Iliad*, for example, begins when Agamemnon incites Achilles's rage by stealing his concubine, Briseis. Agamemnon commits this theft because he has been forced to return his own concubine, Chryseis, to her father Chryses after Chryses prayed to Apollo to send a plague down on the Greek army. As Achilles and Agamemnon argue over their property rights, Briseis's preferences are completely ignored.[36]

Like Herodotus, the Roman historian Livy uses rape as a major narrative device for shaping history. In his 142-book history of all the events "since the city was founded" (*Ab Urbe Condita*), Romulus, the founding father of Rome, is born after the god Mars rapes the Vestal Virgin Rhea Silvia. Livy's description of the event itself is brief: "the Vestal was raped by force" (1.1.8; *vi compressa Vestalis*).[37]

Romulus in turn decides that the best way to ensure population growth in his new city—at the time populated almost exclusively by men—is to invite the neighboring Sabines to Rome, abduct the Sabine women, and force them to bear Roman children (*Ab Urbe Condita* 1.9–13). Finally, the event that sets the downfall of the Roman monarchy and establishment of the Republic in motion is the rape of Lucretia by Sextus Tarquinius (*Ab Urbe Condita* 1.57–58).[38]

The story of Lucretia is particularly gruesome. As Livy tells it, her husband Collatinus brags to a group of men, including Sextus Tarquinius, that his wife Lucretia is more virtuous and chaste than their wives are. The group goes to visit Lucretia and finds her occupied with weaving, unlike their own wives, who were at a luxurious meal. Sextus Tarquinius immediately desires Lucretia, "not only for her beauty, but for her chastity" (1.57.11; *cum forma tum spectata castitas incitat*). A few days later, Sextus Tarquinius goes to Lucretia alone. First, he draws a sword and threatens to kill her if she does not comply; when she refuses, he threatens to kill both her and a slave and make it appear that she had been having an affair. She capitulates, but later tells her husband and father what happened. Lucretia claims that, while her body has been violated, her mind is innocent (*ceterum corpus est tantum violatum, animus insons*); after extracting an oath from Collatinus that he will take vengeance, she commits suicide (1.58.7).

Lucretia, at least, has a clear sense that her consent to the sexual acts that Sextus Tarquinius forced on her was coerced and therefore invalid.[39] Her husband and father believe her and absolve her of guilt. But while consent is crucial to modern definitions of rape, its role in ancient definitions of sexual assault is less clear. The vocabulary used to describe sexual assault in both Greek and Latin map poorly onto our own terminology. Already in this chapter, I have used the words *rape*, *abduction*, and *theft*. This imprecision is due to a corresponding imprecision in both the Greek and Latin languages. In spite of the high frequency of instances of sexual assault in ancient

Greek myth and literature, the Greeks had no word that precisely meant *rape*. Instead, they had several words that approximate our concept of rape, but none of which are exclusively sexual in meaning, including *harpagē* (abduction / theft), *bia* (an act of violence), and *hybris* (a crime or outrageous act, often involving violence).[40]

Rape was not defined in Greece and Rome, as it is in our society, by the lack of or inability to consent to sexual activity. Having sex with an unwilling woman or man was considered problematic in some contexts, but it was not always a crime. Criminality was heavily dependent on the circumstances, especially the intent of the aggressor: was he motivated by a desire to shame and dishonor the woman?[41] Having sex with one's own slave never counted as rape. Stealing a female virgin, on the other hand, *was* a crime— against her father, who suffered a financial loss because virginity was a valuable commodity for procuring a good son-in-law.[42] The rape of a freeborn woman was a crime so severe that it could lead to death for the perpetrator, a sentence far more severe than any in our society. If, however, the man who stole the virgin was unmarried and had sufficient wealth and status, and he offered to marry the victim as restitution, then the crime was offset, because the loss in value was ameliorated.

Such a course of events seems grotesque today. One of the arguments used by those who want to ensure abortion will always be legal for rape victims is that a woman's rapist might somehow get parental rights to the child, effectively retraumatizing the victim by forcing her to interact with her rapist as a co-parent. In ancient Greece and Rome, however, the idea that the rape of a virgin was mitigated if the rapist and victim later got married was not considered quite so horrifying. In fact, such scenarios were common in the plays of the Greek comedian Menander and the Roman comedians Plautus and Terence.[43] Marrying a woman to her rapist seems to have been understood not only as sensible but also as a potentially rich source of humor.

As far as marriage was concerned, in classical Athens—the society of Herodotus and Lysias—the bride's consent was not required. The decision-making power for the bride was entirely in the hands of her *kyrios*, a male relative who had complete legal power over her. She was a passive object in the wedding rather than an active participant, and the marriage contract (*engyēsis*) was merely a transfer of *kyrieia* (guardianship) from the previous *kyrios* to the husband, who then became the new *kyrios*.[44] The previous *kyrios*, however, maintained some control over her dowry.[45] Legally speaking, Athenian women in the classical era had almost no right to consent to marriage, and there was no legal concept of marital rape. The Red Pill community wants to return to this model of *kyrieia* (and its Roman equivalent, *tutela*), and looks to it as the ideal power structure to control female behavior, including the ability to falsely accuse men of rape.

Like the Greeks, the Romans had no word that specifically meant *rape*. Instead, they used a variety of words to describe sexual violence, all of which had additional, non-sexual meanings. *Raptus*, the word from which we get our word for rape, literally means theft. Other words for theft could also be used, such as *iniuria* (insult) and *vitiare* (to defile/corrupt), signifying that, for the Romans, rape was a crime not against the victim but against those who held responsibility for him or her. A Roman woman was almost entirely under the control of the head of her household, the *paterfamilias*. After the death of her *paterfamilias*, the woman was taken into the guardianship of a *tutor* who assumed legal responsibility and control over her; this relationship was called *tutela*. If she was raped, her ability to bear legitimate children was compromised, so the *paterfamilias* or *tutor* could seek punitive measures against the rapist.

Not all Roman rape vocabulary revolved around theft and loss of value. *Comprimo*, the verb used by Livy to describe what Mars does to Rhea Silvia, is often used to describe sexual assault, as is *vis* (force). Like *raptus* and *iniuria*, *vis* could be a criminal charge, similar to what is called forcible rape in America today. *Violare* (to do violence to) has similar implications, and is the word Lucretia uses to describe what has been done to her body. *Stuprum*, a violation, was another criminal charge in the case of acquaintance rape or seduction.

In both Greece and Rome, it was legally impossible for a man to rape his slaves, since they were not granted the privilege of withholding consent.[46] It was also not possible for a husband to rape his wife, yet in Rome there is both legal and literary evidence suggesting that the bride's consent was, surprisingly, a necessary requirement for marriage: Cicero, when considering possible third husbands for his daughter Tullia, rejects one possibility because he thinks Tullia cannot be persuaded to accept him.[47] Legal sources from a few centuries later concur that a legal marriage requires the consent of the bride, the groom, and the fathers of both—if the fathers are still alive, which one scholar estimates would be the case at the first marriages of only half of Roman women of the senatorial class and a third of Roman men.[48] Of course, the definition of consent varies widely, from "enthusiastic agreement" to "lack of opposition" to "agreement under duress," so it is not clear how much the consent of the bride was merely a token gesture.[49]

In a famous scene from Ovid's *Metamorphoses*, Proserpina—more commonly known by her Greek name Persephone—is raped by the god Pluto. When Ceres, Proserpina's mother, voices her outrage to Jupiter about the rape, he responds, "But see things as they stand: Let your words / Fit the facts. Is this a theft (*iniuria*) / Or an act of love? Once you accept him, / This is a son-in-law to be proud of."[50] This tactic—replacing the horror of the event from the perspective of the raped woman with a semantic discussion of the precise vocabulary that ought to be used to describe the actions and motivations of the man, and emphasizing the resulting legal marriage—is a common tactic not only in ancient literature, but also in modern scholarship.

Since the ancient definitions of sexual assault are so difficult to pin down, can the concept of rape even be meaningfully applied to classical antiquity? Some scholars argue that we should stop using euphemisms such as *abduction* and instead "call it rape."[51] Another classicist, Edward Harris, argues *against* using the word *rape* on the basis that we should not impose our modern ideas of rape on the ancient world. He points out that, while sexual violence was widespread in antiquity, the Greeks and Romans responded to different acts that, for us, all fall under the category of rape with a wide range of different emotional responses, from outrage to leniency. As a result, he argues that their concept of what kinds of sexual assault were criminal was so fundamentally different from our own idea of rape that the two cannot be fruitfully compared.[52] He makes this argument especially baldly in his review of the book *Rape in Antiquity: Sexual Violence in the Greek and Roman Worlds*, which he calls "a generally worthwhile collection of essays about a nonexistent topic."[53]

Classical scholars disagree about the degree to which women in ancient Greece and Rome were truly oppressed, and there are many who argue that the social reality of sexual politics would have been far less bleak than the legal evidence leads us to suspect. One may or may not find such arguments convincing; regardless, the point at issue is just how bad it was to be a woman in ancient Athens or Rome. No scholar argues that the overall state of women's rights in the ancient world was worthy of emulation or that our society would benefit from similar laws and customs. But that is precisely the idea that thought leaders in the Red Pill have been espousing recently.

The Death of Hippolytus

The men of the Red Pill believe that we live in a "gynocentric" society, and they argue that the prevalence of false rape allegations and the credibility granted to women who claim they have been assaulted are proof. As Leonid argues in the article which opened this chapter,

> The problem isn't simply that justice isn't being done, it's that the fundamental principles on which justice is founded have been waived in order that the highest source of legal authority—the constitution of our society itself—is the word of women. Such is the Orwellian perfection of this tyranny that people still sincerely believe that we live in a "Patriarchy."[54]

The dichotomy Leonid draws between patriarchy and a society in which rape allegations are given credence is false. Fifth-century BCE Athens and first-century CE Rome were, by almost any reckoning, patriarchal societies, but anxieties about false rape allegations were still very much a concern in those contexts. False allegations therefore cannot be symptoms of gynocentrism. They could, possibly, be a symptom of some deep truth about the female psyche, and evidence that "all women are like that," but they cannot be evidence that men are oppressed by women in our society.

Analyzing these stories does reveal something—not about the inherent insanity or malice of women, but about the male-dominated cultures that produce the alleged rapists. This same approach can usefully be applied to the myth of Phaedra and Hippolytus. A close reading of Phaedra's story shows that her successful false allegation depends upon the fact that she is the "right" kind of accuser with the "right" kind of alleged rapist. She has a perfect record of virtue as defined *by men*; the alleged rapist, Hippolytus, fails to adhere to the standard tenets of masculinity. Ultimately, his punishment is decided by those in power who represent the established norms and who have a vested interest in punishing men who fall outside those norms. Phaedra's false allegation works perfectly within the patriarchal constructs of her society.

Phaedra's story has been told dozens of times, primarily by male authors. Phaedra is the granddaughter of Zeus: after the god abducted Europa, he carried her to Crete, where she gave birth to several children. The eldest, Minos, became king of Crete; he married Pasiphaë, and they had two daughters, Ariadne and Phaedra.

The earliest surviving lengthy treatment of the Phaedra myth, and the one that presents the most sympathetic view of her, is the tragedy *Hippolytus*, written by the Athenian playwright Euripides and first performed in 428 BCE.[55] Euripides's *Hippolytus* begins with an appearance by Aphrodite, the goddess of love. Aphrodite tells the audience that Hippolytus has shown her no honor. He never prays to her, never sacrifices to her, never engages in her rites—that is, he has remained virginal. Because of this insult, Aphrodite has decided to punish him. Her chosen instrument of vengeance is Hippolytus's stepmother Phaedra, whom Aphrodite has caused to fall in love with him.

At the beginning of the play, Phaedra is trying to conceal her suffering and intends to die before revealing her secret passion; she has been starving herself for the past three days, and is now weak and delirious. Aphrodite, however, will not allow Phaedra to succeed in that course of action: "But her passion must not end in that way" (*Hippolytus* 41; ἀλλ᾽ οὔτι ταύτῃ τόνδ᾽ ἔρωτα χρὴ πεσεῖν). While Aphrodite punishes Hippolytus for exercising his agency and choosing *not* to have sex, she denies Phaedra agency entirely.

Worn down by the well-intentioned pleas of the play's chorus and her nurse, a slave companion and motherly figure who has been attending her for years, Phaedra eventually does break her silence and reveal her affliction. The nurse is initially horrified, but then decides—without Phaedra's knowledge or approval—to proposition Hippolytus on her mistress's behalf. His response underscores not only his avowed celibacy, but also overwrites Phaedra as a person with women as a category:

> O Zeus, why did you ever plant women in the light of the
> sun? They are a deceitful evil for mankind. If you wanted
> to perpetuate the human race, you should have made re-
> production happen without women. Instead, men should
> have purchased offspring with a measure of bronze or iron
> or gold, as much as he could afford, and then lived in houses
> free from females. (616–624)

This vision of a woman-free utopia is remarkably similar to the misogynistic fantasies of Hesiod and Semonides, to which I referred in
Chapter 1. Hippolytus almost appears to be an ancient prototype of
the Red Pill's Men Going Their Own Way community, opting out
of the sexual marketplace entirely.

Phaedra overhears his outburst. Her first reaction is fear that if
Hippolytus forswears his oath and tells Theseus what has transpired, Theseus may disinherit their children (715–721). To preemp-
tively discredit Hippolytus, Phaedra hangs herself and leaves a note
for her husband saying that Hippolytus raped her and she was unable to live with the shame. This act is her only exercise of autonomy
in the entire play.

When Theseus finds the note, he is all too ready to believe that
Hippolytus is guilty. Although never stated explicitly, it is heavily
implied that the inveterate womanizer Theseus had long been irri-
tated by his son's outspoken chastity; as a result, he seems perversely
delighted to discover that this performance has been a mask for hypocrisy. Theseus announces, "I give this warning to all: avoid men
like him. They captivate you with their reverent words even as they
plot shameful deeds" (955–957).

Since fidelity is strictly required of wives, Phaedra's previous
chastity is a point in support of her version of the story. But since
men are expected to play the field, Hippolytus's sexual history (or
lack thereof)—his failure to perform masculinity according to social norms—makes others suspicious of him. Hippolytus responds

by trying to explain logically to his father why it would have made no sense for him to rape Phaedra, using arguments such as pointing out that Phaedra was not beautiful enough to be worth the risk: "Did she have the most beautiful body of all women?" (1009–1010; πότερα τὸ τῆσδε σῶμ᾽ ἐκαλλιστεύετο / πασῶν γυναικῶν). Theseus, unconvinced by this well-worn argument, exiles Hippolytus and curses him.[56]

Since Theseus is the son of the god Poseidon, his curses are more than mere words.[57] After a choral ode, a messenger returns to announce that Hippolytus has been in a horrible chariot accident and is close to death. He is brought back onstage, and Theseus is conflicted; although he feels Hippolytus's punishment was more than deserved, he can feel no true joy at the suffering of his son (1257–1260). That ambivalence, however, is erased when the goddess Artemis appears to inform Theseus that Hippolytus was innocent and Phaedra was the liar (1286–1289, 1296–1312). Hippolytus dies in his father's arms, and the play ends.

Euripides's *Hippolytus* is a difficult play to watch. It has no clear hero and no clear villain. Every character is both wrong and wronged. The chorus ends the play by saying, "Floods of tears shall come over us again and again," and that indeed seems to be the only appropriate response to such a painful story with no apparent moral (1464). But the depiction of Phaedra is especially troubling and contradictory. How can someone both make a murderous false rape accusation and still be an essentially good person?[58] Before she is victimized by Aphrodite, Phaedra is a model wife. She never intends to act on her illicit passion, and she never approaches Hippolytus or even interacts with him directly during the play. She is beloved by her nurse and the chorus, who seem genuinely worried about her health and safety.[59] Even Phaedra's final act is intended to protect the legacy of her children.

Later versions of Phaedra's myth depict a character with deeper moral shortcomings than the Phaedra of Euripides's surviving

Hippolytus.[60] In these later myth variants, Phaedra is typically granted more agency: she approaches and propositions Hippolytus, then takes vengeance on him. However, even in these versions, Phaedra's decisions are still heavily restricted by the patriarchy in which she operates. In Euripides's *Hippolytus*, Phaedra's fate is almost entirely decided by Hippolytus's choices; in Roman versions of the story, her father and her husband are the ones whose behavior shapes hers.

Phaedra makes several appearances in the works of Ovid: she is mentioned briefly in the *Ars Amatoria*, and Hippolytus is given a chance to tell his version of the story at more length in *Metamorphoses*.[61] But Ovid also tells the story from Phaedra's perspective in one of his *Heroides*, a work he wrote before the *Ars Amatoria*. The *Heroides* are poetic love letters written by the women of myth to their lovers and a few letters written by the lovers in response. Phaedra's letter to Hippolytus, attempting to verbally seduce him into a relationship, is unusual; most of the poems are accusatory, written at the point in the story when the hero has already abandoned the heroine for a new woman. Perhaps the prototypical letter of the *Heroides* is the letter of Ariadne, Phaedra's sister, to Theseus after he leaves her behind. That episode casts a dark shadow over the Latin versions of the myth of Phaedra and Hippolytus, even though it took place years earlier. Ovid and Seneca after him both link Phaedra's psychology inextricably to her family history.

Theseus is not only Phaedra's husband; he is also the man who abandoned her sister, for no apparent reason, on the remote island of Naxos. This betrayal was especially treacherous after Ariadne played a crucial role in helping Theseus kill the infamous Minotaur, Ariadne and Phaedra's half-brother. Like Phaedra and Ariadne, the Minotaur was the child of Pasiphaë, but he was fathered by a bull after Pasiphaë convinced the craftsman Daedalus to create a mechanical cow costume so she could consummate her lust for the animal.

So Phaedra comes from a family of women known for their doomed passions, and she is married to a man known for his womanizing and faithlessness. These are important factors for the Ovidian and Senecan Phaedras, who have no difficulty dissuading themselves from feeling any guilt for being unfaithful to their husbands. The Phaedra of the *Heroides* even sees the fact that both she and Hippolytus have been treated badly by Theseus as a tenuous bond tying them to each other. "Unless we're going to deny the obvious," she writes to Hippolytus, "Theseus prefers Pirithous to Phaedra and Pirithous to you. Nor is that the only wrong he has done us" (*Heroides* 4.111–113).

Phaedra is not entirely wrong. Pirithous, Theseus's lifetime friend and companion, is the only person to whom the Athenian king can be said to have been faithful. They are constantly paired together, going off on heroic exploits and leaving the women in Theseus's life behind. One of their more famous and troubling adventures is their pact to both sleep with daughters of Zeus / Jupiter. After Theseus accomplishes this heroic feat by abducting and deflowering a barely pubescent Helen from Sparta, the two decide to go down to the underworld so that Pirithous can sleep with Persephone, queen of the dead and wife of the god Hades / Pluto. In Seneca's *Phaedra*, Pluto discovers the plan and detains the two men in the Underworld, leaving Phaedra at home with no husband to deter her from acting on her desire for her stepson.[62]

Seneca's *Phaedra* takes its inspiration from several sources: Ovid's version of the story, Euripides's version, and Seneca's distinctive brand of Stoicism.[63] When we first meet Seneca's Phaedra, she is not starving herself to death as her Euripidean counterpart was. Instead, she is outspoken about her passion for Hippolytus and her lack of loyalty to her faithless husband, currently in the act of displaying what she bitterly calls "his customary fidelity" (92; *quam solet Theseus fidem*).

While Phaedra's nurse in Euripides's *Hippolytus* encourages her mistress to seek sexual relief from Hippolytus, Seneca's nurse acts as a kind of Stoic *proficiens*, advising her as Seneca himself advised Lucilius in his letters: "Extinguish the flames, do not allow yourself to persist in this awful hope: whoever drives love back at the very beginning is always victorious and safe. But once someone nourishes the evil with sweet blandishments and submits to the yoke, it is too late to resist" (*Phaedra* 131–135). Phaedra ignores her nurse's Stoic philosophizing and declares, "I know that what you say is true: but my madness compels me to follow a worse course of action" (177–179). Stoicism emphasizes not assenting to false impressions and responding rationally to situations rather than giving in to strong emotions (*pathē*), so Phaedra's intentional, irrational surrender to lust is anathema to Stoic philosophy in any form.[64]

Seneca's Phaedra approaches Hippolytus and confesses her feelings to him. She feels that falling in love with him was inevitable. Her desire for her young stepson is a self-conscious echo of her older sister Ariadne's desire for a younger version of Theseus (646–666). Hippolytus responds, characteristically, with a long speech decrying Phaedra as even more wicked than her mother Pasiphaë and the rest of her gender. This is his second misogynistic speech of the play; the first he had delivered even before learning about Phaedra's desire for him: "Woman is the beginning of all evil. A fabricator of schemes, she occupies men's minds. Her sexual misdeeds have burned so many cities, led to so many wars, overturned so many kingdoms, oppressed so many people!" (559–564). Although Phaedra changes in different ancient versions of the story, Hippolytus stays stubbornly the same, ready at a moment's notice to extemporize about the immorality of an entire gender with which he has little experience.[65]

The rape accusation in Seneca's play is the nurse's idea, not Phaedra's. But Phaedra takes up the plan enthusiastically. And unlike Euripides's Phaedra, who leaves a rape accusation in a suicide note, Seneca's Phaedra contrives to have physical proof. During their confrontation, Hippolytus draws his sword; when she holds it to her breast, encouraging him to kill her, he throws away the sword in disgust (704–712). Phaedra then uses the sword as a prop, telling Theseus that Hippolytus left it behind after he violently raped her (896–897). After hearing this story, Theseus exercises one of his three wishes from his father Neptune to curse Hippolytus. After Hippolytus dies, Phaedra returns to the stage to hysterically confess her guilt before stabbing herself with the sword she used earlier as false proof (1191–1200).

In every version of the story, Theseus believes Phaedra completely. Nobody asks or speculates about whether fasting made her memories less reliable, or what Phaedra was wearing when Hippolytus allegedly raped her. Because Phaedra's deportment has always been so flawless, she is the rare woman who is likely to be believed, so her false accusation has serious consequences for the accused.

Before her suicide, Phaedra is an exemplary woman when judged by the patriarchal society in which she lived. She has even internalized the pressure to conform to social norms for female behavior. In Euripides's *Hippolytus*, Phaedra is obsessed with her *aidos*, her shame. She has worked hard to maintain her good name and flawless reputation, and if her feelings for Hippolytus are discovered, that carefully erected *aidos* will be torn to pieces. But what is shame other than a negative emotional response used to punish someone who fails to conform to social norms? In a monumental study of *aidos* in Greek literature, Douglas Cairns defines it as "that feeling of inhibition which holds one back from socially unacceptable action."[66] When applied to Phaedra, her *aidos* is a manifestation of how successful she has been at following prescriptions for female behavior as set down by men. She even calls *aidos* a pleasure (*hēdonē*), a formulation that has puzzled many scholars—but the idea that a woman might take pleasure in having flawlessly conformed to the patriarchal ideal of wifehood is not difficult to comprehend (383–387).[67]

Phaedra's older sister Ariadne is the subversive one. Ariadne is the sister who chose Theseus as her husband and betrayed her father for him. But Ariadne was punished for that act by Theseus's abandonment. Phaedra, unlike her sister, is married off by her father to Theseus—possibly against her will, as Seneca's version of Phaedra explicitly says in her first speech (lines 89–91). But she, too, was left entirely at the whims of a philandering and cruel husband. Her passion and her subsequent false accusation are both created and shaped by a deeply chauvinistic society.[68]

Red Pill Marriage

False rape allegations are often used by Red Pill writers as the ultimate proof that we live in a gynocentric, misandrist society. But analyzing the narratives around false allegations, both ancient and modern, shows that the opposite is true. Susan Brownmiller's groundbreaking feminist analysis of rape myths, defined as "prejudicial, stereotypical and false beliefs about rape, rape victims and rapists"—such as the idea that false accusations are common, or that women secretly want to be raped, or that rape can be enjoyable—showed how rape myths are used to perpetuate rape culture.[69] As Colleen Ward puts it, building on Brownmiller twenty years later,

> rape myths in patriarchal societies underpin sexual violence and are responsible for the unjust treatment of women who have been victims of sexual assault. Men, as oppressors of women, are more likely to cling to rape myths and are more likely to blame and denigrate victims of sexual abuse. This allows men to retain authority and control and constrains women to positions of powerlessness in society.[70]

But we can go even further. Not only is the belief that false accusations are widespread the product of a patriarchal society; the rare instances in which they actually do occur are *also* the product of patriarchal societies.

On December 7, 2014, Daryush Valizadeh made the following claim in his video "All Public Rape Allegations Are False":

> I'm calling it right now, that all public rape accusations whereby the woman went to the media or called the media tip line before reporting the crime to the police is [*sic*] false. Any time a woman publicly broadcasts a rape before using the proper channels, before going to the authorities, means that she is a liar and in fact was not raped.[71]

Valizadeh scripts the appropriate way for a rape victim to act. She must immediately put herself in the hands of a legal system that was created by male lawmakers and is enforced by a predominantly male police force (88.4 percent male, according to a 2013 FBI study). If she does not act in a manner that places justice in the hands of these male structures of authority, "she is a liar and in fact was not raped."[72]

Valizadeh is utilizing an extremely specific straw man. There is little logic to or evidence for the idea that a woman would call up the media with a random, unverifiable claim and expect to get massive amounts of attention without close scrutiny, as Valizadeh implies. But that media straw man provides the foundation for the much more general and pernicious statement he concludes with, "When a woman cries rape, look at the evidence, look at the conviction that happens in a court of law where the man was given due process. But other than that, she is full of shit. So, that's all I wanted to say. Be skeptical."[73]

The reasoning of those last few sentences is absurd. Valizadeh seems to think that he is suggesting something revolutionary, as though he were the lone voice of reason fighting against the default response of automatically believing all rape allegations—when really, the opposite is true, and the stance of "believing the victims" is far from mainstream. Catharine MacKinnon argued thirty years ago that "feminism is built on believing women's accounts of sexual use and abuse by men," but as recently as 2014, Jessica Valenti's assertion that we ought to "start to believe victims en masse" was called "particularly appalling" in an article in *Time* because "undermining the presumption of innocence is not good for women, either."[74] The mere existence of Valenti's article and the backlash to it show that our willingness to trust victims is not what Valizadeh implies it is. Furthermore, the idea that not enough concern is given to the impact on the falsely accused men is ludicrous when judges worry about ruining the lives even of young white men who are *actually convicted* of rape—where there is, as Valizadeh requests, a "conviction that happens in a court of law where the man was given due process."[75]

The men of the Red Pill community are predominantly white, and although a false rape allegation can certainly ruin a white man's reputation, it is extremely unlikely to lead to his being incarcerated or even indicted. If the accused rapist is white, many people's first impulse is to disbelieve the credibility of the narrative altogether and instead question the victim's memory or behavior. Even when we are willing to accept that sexual activity took place, the assumption is that the woman offered some kind of tacit consent by dressing provocatively or not saying "no" firmly enough. One article about false allegations points out that women and children are "defined as trusting and yet, in the sexual sphere, not to be trust*ed*."[76]

From a feminist perspective, the most remarkable and unlikely aspect of Phaedra's story is that Theseus believes her and acts immediately and decisively on that belief. But her false accusation does, as Valizadeh stipulates it ought, use "the proper channels": she reports the false crime to her husband Theseus, the king, who then calls on the god Poseidon for supernatural vengeance.[77]

The society of Theseus, Phaedra, and Hippolytus is not just mythical to the men of the Red Pill: it is also aspirational. They look both backward and forward to a time when, from a feminist perspective, all heterosexual sex would be rape—and, from a patriarchal perspective, almost no sex would be. Men in all factions of the Red Pill have become increasingly outspoken in recent years about their desire to bring back not just old-fashioned but literally antiquated gender politics—to reinstate the social conditions of men and women in the ancient world, where it would be impossible to disagree with Dworkin that intercourse would occur in a context where "men have social, economic, political, and physical power over women."

Earlier in this chapter, I mentioned that Valizadeh has been advocating since 2015 for the idea that "women must have their behavior and decisions controlled by men"—an idea with obvious similarities to ancient models in which women lived under the guidance of a *kyrios* or *paterfamilias*. He proposes

> two different options for protecting women from their obviously deficient decision making. The first is to have a designated male guardian give approval on all decisions that affect her well-being. Such a guardian should be her father by default, but in the case a father is absent, another male relative can be appointed. . . .
>
> She must seek approval by her guardian concerning diet, education, boyfriends, travel, friends, entertainment, exercise regime, marriage, and appearance, including choice of clothing. A woman must get a green light from her guardian before having sex with any man, before wearing a certain outfit, before coloring her hair green, and

before going to a Spanish island for the summer with her
female friends.

If she disobeys her guardian, an escalating series of punishments would be served to her, culminating in full-time
supervision by him. Once the woman is married, her husband will gradually take over guardian duties, and strictly
monitor his wife's behavior.[78]

The second proposal is for "a combination of rigid cultural rules
and sex-specific laws," reminiscent of the *aidos* that Phaedra has so
internalized in Euripides's *Hippolytus*.

Valizadeh's justification for this argument seems to draw from
ideas he internalized through the study of ancient Stoic texts:

I make these sincere recommendations *not out of anger*, but
under the firm belief that the lives of my female relatives
would certainly be better tomorrow if they were required
to get my approval before making any decisions. They
would not like it, surely, but *due to the fact that I'm male
and they're not, my analytical decision-making faculty is superior to theirs* to [sic] absolutely no fault of their own,
meaning that their most sincere attempts to make good
decisions will have a failure rate larger than if I was able to
make those decisions for them.[79]

The assertions that he is not motivated by anger and that men are
more analytical and less emotional than women are both validated
by Red Pill Stoicism. As I argued in Chapter 2, the Red Pill community reads ancient Stoic texts in large part to perpetuate their
belief that they alone are rational enough to understand the world
unemotionally, and therefore they should be in charge. The appeal
of dead white male intellectual giants also plays a role in that interest, and Valizadeh subtly reminds his reader of their legacy when

he ends his article with this alarmist declaration: "Unless we take action soon to reconsider the freedoms that women now have, *the very survival of Western civilization is at stake.*"[80]

Since the publication of that piece in 2015, Valizadeh has moved on to more concrete policy suggestions. In the blog post "How to Save Western Civilization," he argues that "we must repeal women's suffrage, starting with the 19th Amendment in the United States. Once this is accomplished, no other planned or conscious action must be taken to solve nearly all our societal ills." He claims that, if only men were allowed to vote, they would enact pro-male laws that would be for the benefit of everyone, beginning a steady trend toward an increasingly patriarchal society that would improve the lives of all.[81]

Ideas about race are coded into Valizadeh's arguments, but rarely expressed explicitly. His fixation on the Nineteenth Amendment is telling—although it prohibited citizens from being denied the right to vote on account of their gender, women of color were still denied voting rights based on their race for decades after the amendment's ratification, until the Voting Rights Act of 1965 overruled the Jim Crow laws restricting the civil rights of people of color. But by fixating on 1920 and the Nineteenth Amendment as the moment when the U.S. took a wrong turn, Valizadeh reveals that his ideas are centered around the voting rights and behavior of white women. Although Valizadeh does not state this outright, that implication is clear from his claims about the relationship between controlling female behavior and the future of "Western civilization," the latter of which is the primary interest of the so-called Alt-Right.

Valizadeh is insufficiently white to be embraced by the white supremacists of the most extreme parts of the Alt-Right, but their ideas are similar to his. Rather than looking to ancient models, as Valizadeh does, neo-Nazi Andrew Anglin is now advocating for what he calls "White Sharia." When an image surfaced of Nathan Damigo, the head of the white nationalist group Identity Evropa, punching a female antifascist protester, Anglin responded with an article celebrating Damigo for embodying "WHITE SHARIA in ACTION!" After doxxing the protester, Anglin shared images cataloguing her descent from, in his words, "nice, sweet little girl" to "hairy fetish model." He laments, "When this bitch was 15, you could genuinely say 'Aryan princess.' She was glowingly pretty, I'm sure very sweet, giggly, would blush if you flirted with her . . . but 'freedom' and the university system turned her into a hairy porno monster rioting on the street." He continues,

> The degree to which you allow women freedom is the degree to which they will destroy their own lives and the lives of everyone around them and *by extension all of civilization* itself. . . .
>
> We have to forcibly marry-off [*sic*] teenage girls to men who will TAKE CARE OF THEM!
>
> We WANT women to be taken care of, as they are the thing that cooks, cleans, gives sexual pleasure and produces and raises children! The white knights don't care about the protection of women! White knights skirt their MASCULINE responsibility by refusing their biological duty to protect women in REAL terms, which mainly involves protecting them from themselves, rather than some external threat![82]

When Anglin writes about "women," he really means "white women." It is their autonomy that, to him, constitutes a threat to our very society.

Similar arguments have also appeared recently on *Chateau Heartiste*, a website that combines seduction advice and alt-right ideas. In the article "Single White Women Want to Spread Their Legs for the World," James "Roissy" Weidmann argues that white women cannot control their irresistible attraction to men of color.

He concludes that we must "rescind suffrage and disenfranchise single White women" and "get more White women married off and pregnant at younger ages."[83] Here, the connection between these writers' desire to control women and their racial anxieties is made plain.

All three of these prominent Red Pill writers—Valizadeh, Anglin, and Weidmann—link female freedom (and especially sexual freedom) to the downfall of society. The only way to save Western civilization is to restrict female behavior, and particularly white female reproduction: men must be the ones to decide for women at what age and with whom they will have sex. These men are not concerned about female consent, because they believe that what women want is usually bad for them. They argue that paternalistic control of the type that existed in classical antiquity is absolutely necessary to create a society that will be good for both men and women.

These ideas can also be found in one of the darkest corners of the Red Pill world: the community of "incels" (short for involuntarily celibate) whose sexual failures have filled them with toxic, occasionally murderous rage at women as a collective.[84] Discussions of the value of "forced marriage" are common on incel fora such as the r/incel subreddit, which was shut down by the site in late 2017 for promoting violence against women. Incels often claim that, because women have thus far chosen not to have sex with them, women should not be allowed to choose their sexual partners at all.

In this Red Pill utopia, the definition of rape would be extremely narrow—but, from a feminist perspective, almost all sexual activity would occur under highly coercive conditions. Rape in such a society would not be a crime against the woman herself: since her sexual choice would be entirely under the jurisdiction of her male guardian (a father or husband, usually), rape would be a failure to obtain *his* consent, not hers. Her consent would be almost immaterial (although not entirely so, as the story of Lucretia suggests). The sexual politics of this Red Pill utopia would be eerily similar to those of classical Athens and Rome under the late Republic and early Empire. In fact, wives in Rome may even have lived under preferable circumstances: Red Pill writers never clarify whether women would be allowed to own property and whether their consent would need to be obtained before entering into a marriage.

This is the context in which all discussion of rape and false rape in the Red Pill must be read. In this imagined ideal future, false rape allegations will still be possible under extremely limited circumstances: the story of Phaedra and Hippolytus proves that such scenarios can still exist even when women are controlled by men. The Phaedra myth does not work for the Red Pill as an accurate reflection of the world in which we live: it is a worst-case scenario for the world in which they *wish* we lived.

———————

As I have argued, close reading of Phaedra's story reveals that the tragedy owes far more to the behavior of her male relatives and the patriarchal society in which she lives than to any flaw in her character or, for that matter, female psychology in general. Nevertheless, there are feminist scholars who disapprove of her.

In May 2015, the prominent British classicist Edith Hall published a post, "Why I Hate the Myth of Phaedra and Hippolytus," on her blog *The Edithorial*. It is worth quoting at length, because Hall summarizes many aspects of the feminist position I find problematic concerning false rape accusations:

> On Friday Hampshire Police finally apologised for mistreating a rape victim from Winchester. She reported the crime at the age of 17 in 2012. They threatened her with prosecution for lying about the attack. I remember a male "friend" crowing to me in 2013 that she was one of the allegedly enormous number of women who, "like Phaedra,"

frame innocent men for sex crimes because they have been rejected or out of simple spite.

When the police finally bothered to do forensic texts [*sic*] on the T-shirt she had provided, they realised her evidence was entirely credible. The rapist was charged and convicted. The case has clarified my intuitive loathing of Euripides' tragedy *Hippolytus*, a play of exquisite poetic beauty but toxic ideology in which Hippolytus' stepmother Phaedra falsely accuses him of rape.

Between the Greek original, Seneca's *Phaedra* and Racine's *Phèdre*, let alone descendants like O'Neill's *Desire under the Elms* and Mike Nichols' *The Graduate*, this story has been mightily applauded on centuries of stages and screens. Countless star actresses like Bernhardt and Mirren crave playing the mendacious rape-accuser of fiction. Every performance constitutes another "proof" of the mass delusion that information imparted by women is unreliable—the delusion which philosopher Miranda Fricker calls *Epistemic Injustice* against them.[85]

Hall does not go so far as to say that we should stop studying or performing these plays, or that actors who play the role of Phaedra are responsible for rape victims' not being believed—but the implication is clear. As compelling as the character of Phaedra is, and in spite of the "exquisite poetic beauty" of *Hippolytus*, she believes women would be believed more readily if the play did not exist.

Hall is not the first person to make this claim. The idea that Phaedra harms female credibility goes back thousands of years, to just a few decades after the original performance of Euripides's play. In 411 BCE, Euripides's contemporary Aristophanes put on the comedy *Thesmophoriazusae*. In that play, the women of Athens have gathered together for the annual Thesmophoria, a festival celebration exclusively for women. They take advantage of the gathering to hold a trial and decide whether to execute Euripides, their grievance being that he "portrays women badly" (κακῶς αὐτὰς λέγω) in his plays (*Thesmo*. 85). As a result, Athenian husbands have become much more suspicious, and Athenian wives are less able to fake pregnancies, steal food, and meet with their lovers.

Phaedra is the primary example given in the play of an evil Euripidean heroine.[86] She is brought up repeatedly throughout the play. The first references to her are by Euripides's male relative (in disguise as a woman), who jokes early in the play that Phaedra must have preferred to be on top during sex, perhaps because she was a sexual aggressor (*Thesmo*. 153). Later, he also attempts to defend Euripides to the women of Athens by saying, "If he reviles Phaedra, what does it matter to us?" (*Thesmo*. 497–498).[87] The women seize on this example of Phaedra as the paradigmatic evil heroine:

> **Mica:** Shouldn't you be punished? You're the only woman who has dared to speak in defense of the man who has done us wrong, choosing plots with wicked women. He creates Melanippes and Phaedras, but he has never created a Penelope, since she's known for being chaste.
>
> **Relative:** I know the reason for that. You couldn't call any living woman a Penelope—we're all Phaedras. (*Thesmo*. 544–548)

The point made by the women in Aristophanes's comedy from 411 BCE is functionally identical to the one made by Edith Hall in 2015: "Every performance [of *Hippolytus*] constitutes another 'proof' of the mass delusion that information imparted by women is unreliable." The myth of Phaedra, some believe, has the power to influence the way men think about female behavior in the real world.

But what Phaedra's story actually shows is that male terror of false rape allegations is and was largely unfounded. The only way an allegation will be given weight is if the right kind of false victim accuses someone whom others agree acts like a rapist, then provides some tangible evidence to support her claim. Otherwise, the chances of her being taken seriously are nearly nonexistent. In some ways, not much has changed in the last twenty-five hundred years: their anxieties about female sexuality and credibility resemble ours to a surprising degree.

The men of the Red Pill claim that, in a truly just world, there would be no Phaedras. But in the world they idealize, women would truly, as Euripides's relative asserts in the *Thesmophoriazusae, all* be Phaedras. Any time a woman claimed that a man had sex with her against her will, the matter would become a civil dispute between her male guardian and the accused rapist. Her rights over her sexuality and reproduction would be restricted so narrowly that her ability to consent would become essentially meaningless. There would be far fewer false rape allegations—and, for that matter, far fewer true rape allegations. The Red Pill solution to the false allegation problem is to create a society in which female consent is virtually meaningless, so that when a woman asserts that she has not consented, few will care.

It is extremely unlikely, of course, that the fantasies of the Red Pill will become reality. Their dreams would require a complete restructuring of our society's sexual politics: in Weidmann's post on *Chateau Heartiste* about white women's uncontrollable sexual attraction to people of color, he admits, "There aren't many solutions to this intractable cognitive block in women's hindbrains that don't require serious divestment from the recently operative political and social calculus."[88] Realistically speaking, their numbers are too few and their views too extreme for the Red Pill to have any significant political impact. American women need not fear that the ancient customs of *kyrieia* or *tutela* will return.

Practicality aside, however, those same heavily restrictive ancient customs are more than merely attractive to the men of the Red

Pill: they are aspirational. Even more disturbingly, their online vitriol suggests that they believe they *deserve* such a society, and that our society—one in which women are allowed to vote and to choose with whom to have sex and whom to marry—represents a deep systemic injustice against white men. They use ancient literature to justify their sense of entitlement to female bodies and to political power over them.

CONCLUSION

The men of the Red Pill use and abuse classical antiquity in a variety of ways. They have found ample material from ancient Greece and Rome to support their ideology, from Stoic self-help manuals to Ovidian seduction advice to ancient models of patriarchal marriage. Although their analyses of ancient sources rarely display much understanding of context and nuance, Red Pill writers nevertheless are adept at manipulating ancient sources to make them speak meaningfully to contemporary concerns. They have appropriated the texts and history of ancient Greece and Rome to bolster their most abhorrent ideas: that all women are deceitful and degenerate; that white men are by nature more rational than (and therefore superior to) everyone else; that women's sexual boundaries exist to be manipulated and crossed; and, finally, that society as a whole would benefit if men were given the responsibility for making all decisions for women, particularly over their sexual and reproductive choices.

But the Greek and Roman Classics are not beloved only by the radicalized young men of the Alt-Right and the manosphere. They are also claimed by many feminists, including myself. Classical literature and history can be challenging to study as a feminist: female voices from the ancient world are largely silent, while there is no shortage of male-authored narratives of misogyny, abuse, and sexual assault. Nevertheless, there is a long and rich history of brilliant progressive and feminist thinkers using ancient Greece and

Rome to understand patriarchy, oppression, and resistance, from Simone de Beauvoir to Judith Butler and beyond. And there is also a fascinating tradition of the subversive use of the symbols of antiquity to challenge established ideas of cultural value, as when Jay-Z quotes Plato's *Euthyphro*.[1]

The men of the Red Pill are unwilling to accept that those with liberal political beliefs can also appreciate classical antiquity. I experienced this disbelief firsthand when I wrote an article encouraging professional classicists to take note of and engage with far-right misappropriation of ancient texts.[2] I published "How to Be a Good Classicist under a Bad Emperor" in November 2016, at a strange moment in this book's history. I had submitted my first manuscript on November 3, less than a week before the 2016 presidential election—an election I was certain Hillary Clinton would win. Because of that certainty, I put a great deal of energy in that first draft into convincing the reader that the Red Pill community, including the Alt-Right, was worth paying attention to, even though I believed that for most readers the Alt-Right would soon be a distant, painful memory. After the election, my vision for what this book needed to accomplish changed, and "How to Be a Good Classicist" was part of that re-imagination process. I argued that classicists needed to take steps to ensure that the discipline's future would not resemble its past—that we would not return to a world where the Classics were read and championed almost exclusively by wealthy white men.

Predictably, I became the target of a troll storm, and in the days and weeks following, I received hundreds of anti-Semitic tweets and emails, many with attached images of my face (and those of my family members) photoshopped into a gas chamber or a concentration camp. Others threatened me with sexual assault or detailed which gun they would like to use to shoot me. Daryush Valizadeh bragged to his followers that he knew where I and my family lived, but argued that no physical violence was necessary because he had already raped my mind. I still receive messages with anti-Semitic slurs on a weekly basis.

That response, alarming as it was, did not surprise me as much as the more substantive response articles accusing me of hating and wanting to destroy the Classics. One writer wrote, "Had Zuckerberg her way, the ancient wisdom of Athens and Rome would likewise be consigned to the flames." Another speculated, "I think that deep down, you despise these books."[3] The idea of a feminist who enjoys and finds meaning in studying the ancient world is so inconceivable to these writers that they can more easily believe I have spent over a decade studying material I secretly despise.

This disbelief is illuminating. It shows that these men enjoy learning about the ancient world because they believe that they and the ancients share similar beliefs, and they think that anybody who is not as misogynistic and xenophobic as the Greeks and Romans must not truly appreciate and wish to preserve ancient literature and culture. They believe that only a pickup artist can truly appreciate the *Ars Amatoria* and only a Man Going His Own Way can comprehend the anti-marriage screeds of Hesiod and Semonides. According to some, the texts that lie at the foundation of white supremacy and patriarchy can only be enjoyed and understood by those who wish to perpetuate and strengthen those same oppressive structures today.

This is an idea that we ought to take seriously, because it is not far removed from the attitude of some on the political left. The men of the Red Pill believe that the Classics are only (or at least especially) meaningful to reactionary white men, and those with progressive politics who seek to upend or replace the Western canon tacitly cede this point. Both sides of that debate agree that the study of ancient literature perpetuates white male supremacy; they differ only on the question of whether that is a consequence that should be celebrated.

If I agreed that progressives should refrain from studying texts written by dead white men with views that I find problematic, I could not have researched and written this book. I believe it is more useful to analyze *why* the Red Pill community feels so comfortable seeing itself reflected in the ancient world, and more useful to accept that progressives may not feel quite so at ease when studying ancient literature and history. That discomfort is not something to be avoided. It must be embraced as a necessary and productive component of thoughtful, ethical classical reception for today's world.

The vehemence of the response to my November 2016 article shows how much these men fear that a sophisticated, liberal version of classical studies will undermine their self-presentation as the inheritors and protectors of the classical tradition. To them, the Classics need to be protected from everyone with progressive politics, because progressive politics and appreciation of the ancient world cannot coexist. The idea of a vibrant, radical, intersectional feminist Classics—one that uses the ancient world to enrich conversations about race, gender, and social justice—is anathema to them. And that is why feminist Classics today is more exciting and necessary than ever.

The future of the Red Pill community is unclear, but the use of ancient Greece and Rome by these far-right online groups are clues to what the future of Classics may look like and could look like. Classical scholars must accept that, in the twenty-first century, some of the most controversial and consequential discussions about the legacy of ancient Greece and Rome are happening not in the conventional realms of literature, theater, and scholarship, but on the internet. This is not bad news; Red Pill websites have conclusively proven that the ancient world remains a valuable toolkit for thinking through the issues and concerns that plague us today. I do not agree with their ideology or their methodology, but I am in agreement with them that ancient Greece and Rome remain highly relevant to the modern world.

The Red Pill has made going online and voicing opinions perilous for women like me. But in spite of their shaming, trolling, threats, and abuse, they have not succeeded at forcing us off the internet. Nor, despite their use of the ancient world to condone their antiquated gender norms, will they succeed at stopping feminists from enjoying and admiring the Stoics and Ovid.

Feminist Classics in the age of digital misogyny cannot ignore how easily the ancient world lends itself to use by communities such as the Red Pill. However, Red Pill Classics tends to be relatively superficial, relying on the uninterrogated assumption that its members are the natural inheritors of the legacy of classical antiquity. Critical analysis of their use of the ancient world exposes how their classical references work as a powerful rhetorical tool; it also reveals how nuanced, feminist interpretation of the Classics can counteract Red Pill distortions. That is what I hope I have achieved in this book.

Feminists deserve a better internet. And future generations of readers deserve a better kind of discourse about the ancient world: one that is free of elitism and neither uncritically admiring nor rashly dismissive.

GLOSSARY OF RED PILL TERMS

AFC	"Average Frustrated Chump," a man who is unsuccessful with women.
alpha male	A man who is socially dominant and attractive to women.
Alt-Right	"Alternative Right," a neoreactionary white nationalist movement.
AMOG	"Alpha Male of the Group."
ASD	"Anti-Slut Defense," female strategies erected to keep from seeming sexually promiscuous.
AVFM	A Voice for Men, the primary internet hub for the MHRM.
AWALT	"All Women Are Like That"; see also NAWALT.
carousel	Also known as "the cock carousel," a period in a woman's life, usually in her early twenties, when she is extremely promiscuous before settling down.
DHV	"Demonstration of High Value," any behavior meant to prove the high SMV of a PUA.
DLV	"Demonstration of Low Value," any behavior that decreases the perceived SMV of a PUA, including showing too much interest in a target.
frame	A subject's perception of the situation. If he attempts to influence others to accept his frame, he exerts "frame control."

GamerGate	A movement to lessen the impact of social justice movement in the video game and science fiction communities.
gynocentrism	A social order that predominantly focuses on women, to the exclusion of male interests.
hamstering	Rationalizing otherwise indefensible behavior, so called because the "rationalization hamster" spins in its wheel while the woman attempts to figure out a reasonable excuse.
HB	"Hot Babe / Bitch," often followed by a number from 1–10.
hypergamy	A female tendency to always attempt to become the partner of the highest status male, also known as "gold-digging."
incel	"Involuntary Celibate."
IOI	"Indicator of Interest," signs that PUAs look for to determine whether a target is receptive.
kino	Kinesthetics, physical touching used by a PUA to escalate attraction.
LMR	"Last-Minute Resistance," an ASD tactic used to stop sexual intercourse at the last possible moment.
LTR	"Long-Term Relationship."
mangina	A feminist male or "white knight" who puts female interests ahead of his own.
manosphere	The set of internet communities focused on the interests of men, particularly the MHRM, PUA, and MGTOW communities. Adapted from *blogosphere*.
meme	An endlessly replicable and malleable unit, often an image or an image / text combination, used to communicate concepts in shorthand online.

MGTOW	"Men Going Their Own Way," a faction within the manosphere focused on living independently of women, up to the point of avoiding marriage.
MHRM	"Men's Human Rights Movement," a faction within the manosphere dedicated primarily to fathers' rights and fighting false rape allegations.
MRA *or* MHRA	"Men's Rights Activist" / "Men's Human Rights Advocate," a man who is part of the MHRM.
mudsharking	The sexual pursuit of black men by white women.
Mystery	Erik von Markovik, author of *The Mystery Method* and a main character in Neil Strauss' memoir *The Game: Penetrating the Secret Society of Pickup Artists*.
NAWALT	"Not All Women Are Like That," a form of exceptionalism eschewed by the manosphere.
neg	A backhanded compliment used by a PUA to decrease the target's perception of her own SMV.
Obsidian	Mumia Ali, a black MHRA / PUA.
PUA	Pickup artist, a man who has studied the art of seduction.
Quintus Curtius	Writer for *Return of Kings* who focuses on history, philosophy, and Great Men.
r / theredpill	A subreddit dedicated to discussing "red pill" concepts.
Return of Kings	One of the most misogynistic websites in the manosphere, run by Daryush "Roosh V" Valizadeh.
Roissy	James Weidmann, creator of the Alt-Right and PUA blog *Chateau Heartiste*.

Roosh V	Daryush Valizadeh, proprietor of the site Return of Kings and the Roosh V Forums, author of *Bang, Day Bang, 30 Bangs*, and other similarly titled volumes.
SJW	"Social Justice Warrior."
SMV	"Sexual Marketplace Value."
subreddit	A forum on Reddit dedicated to a particular topic.
troll	Someone who posts provocative material online with the aim of provoking outrage and controversy and disrupting discourse.
troll storm	Collective action by a group of trolls to harass a specific target by flooding their social media activity with offensive content.
TRP	"The Red Pill," a central concept in the manosphere (adopted from the film *The Matrix*): recognition of the politically incorrect truth that society is unfair to men, particularly heterosexual white men, and is designed to favor women.
Vox Day	Theodore Beale, science fiction writer and proprietor of the blogs *Vox Popoli* and *Alpha Game*.
white knights	Men who have not "swallowed the red pill" and seek to protect and nurture undeserving women.
ZFG	"Zero Fucks Given."

NOTES

Introduction

1. On the use of ancient sculpture to perpetuate white supremacy in the United States, see Bond 2017.

2. The classic study of the use of ancient Greece in Nazi Germany is Losemann (1977); Marchand (1996) also studies the legacy and roots of German philhellenism. On Hitler's use of Rome and Sparta as models, see Losemann 2007, 308–311 and Roche 2013. Fascist classical appropriation is also a concern in the twenty-first century, not only in the American neo-Nazi movement but also in the fascist Golden Dawn (Χρυσή Αυγή) political party in Greece; see Hanink 2017, 243–250.

3. The article, "What Donald Trump's Victory Means for Men," was written by Daryush "Roosh V" Valizadeh on his website *Return of Kings* just a few days after the election (Valizadeh 2016b). I return to Valizadeh and *Return of Kings* throughout the book.

4. Weiner 2016.

5. Anton's identity was revealed by the *Weekly Standard* in February 2017.

6. Both men also share an interest in the fifth-century BCE Greek historian Thucydides: see Crowley 2017.

7. Even within the so-called Alt-Right, a white nationalist movement, there is a schism with what is sometimes called the Alt-Light, a movement more focused on Western culture and civilization rather than explicitly racial politics. Chapter 1 looks further at this distinction.

8. Painter (2010) answers the question "Were there 'white' people in antiquity?" with, "No, for neither the idea of race nor the idea of 'white' people had

been invented, and people's skin color did not carry useful meaning." On the invention of whiteness and classical antiquity, see Dee 2003–4 and McCoskey 2012.

9. For example, see Sims (2010), an article in the *American Renaissance*—which is published by the notorious white supremacist Jared Taylor—that cherry-picks examples from ancient literature to show that the elites of both ancient Greece and Rome were of "Nordic" descent. Another telling example of racialized undertones in far-right classical appropriation is various gun groups' use of the phrase *molōn labe* (come and get them), which Plutarch attributes to the Spartan king Leonidas in response to Xerxes's demand that the Spartans hand over their weapons (*Apophthegmata Laconica* 225, C11–12). Although the ancient Persians were, of course, not Muslim, Leonidas seems to hold special appeal for Islamophobic gun enthusiasts.

10. For a sociological approach to the community that relies on the methods of digital anthropology, see L. Kendall 2002 and Schmitz and Kazak 2016.

11. Nagle 2017. Nagle's book is a very useful and lively introduction to the political and ideological forces that animate the Red Pill movement, particularly in its "Gramscians of the Alt-Right" chapter (39–52). However, Nagle frequently seems to draw her authority to comment on the Alt-Right from her willingness to criticize the Left: her arguments often cede that the Alt-Right's resentments and anger come from reasonable roots but are taken to unreasonable extremes. Among the progressive objects of her disdain are feminist video games, call-out culture on Tumblr, trigger warnings, and protests against far-right speakers (21, 69–80, 80, 119).

12. This is the argument made by Beard (2017), which concludes, "we probably don't need to worry too much about these alt-Right guys hi-jacking Classics if they make such a mess of it. We just need to keep on pointing out the howlers." I articulate my argument against this type of response more fully in Zuckerberg 2017.

1   Arms and the Manosphere

1. Zvan 2014. Some outspoken figureheads of the community are not heterosexual, including Milo Yiannopoulos and "gay masculinist" Jack Donovan: see O'Connor 2017. However, these men are far from universally accepted

within the larger Red Pill sphere and are often either tokenized or labeled "degenerates."

2. Another example are the *kakuhido* in Japan, a group that combines Marxism and men's rights by, for example, protesting the "passion capitalism" of Valentine's Day.

3. Kupers 2005, 714. Another classic interpretation of authoritarianism and manhood is Theweleit (1977), brilliantly reviewed in light of the Trump campaign in Schambelan (2016).

4. Wachowski and Wachowski 1999.

5. For example, Schmitz and Kazak (2016) divide the community into two subgroups with distinct ideological strategies: "Cyber Lads in Search of Masculinity" and "Virtual Victims in Search of Equality."

6. These men have largely chosen *not* to use emic designations for social justice groups: they regularly refer to transgender people as *trannies*, and they have come up with their own names for feminist websites, such as *Jizzebel* for *Jezebel*.

7. See, for example, Daniels (2009) for a study of white supremacy on the internet in the early twenty-first century, particularly on sites such as *Storm-front*, before the "Alternative Right" existed in its current form.

8. Fisher served in the New Hampshire House of Representatives from 2014 until he resigned in May 2017, after he was unmasked in Bacarisse (2017). After the article was published, Fisher released a statement admitting he had written many of the quotes attributed to him by Bacarisse but claiming that the quotes were taken out of context.

9. Elam 2012.

10. Rensin 2015. See *contra* Serwer and Baker (2015) arguing that Elam's "advocacy efforts are difficult to discern" and that Elam used his history as a deadbeat dad to create a business.

11. This similarity is not a complete surprise. In the 1970s, Warren Farrell, the intellectual grandfather of the men's rights movement, was an associate of Gloria Steinem and Barbara Walters and a fierce advocate for demolishing gender roles that he saw as destructive to both sexes. His views gradually became more radical, culminating in the 1993 publication of *The Myth of Male Power: Why Men Are the Disposable Sex*, the book that caused Paul Elam's "red pill" moment. See Blake 2015.

12. The MHRM is also unique in its similarity to and alliances with men's rights and fathers' rights movements in other countries. A men's rights movement in India, for example, has also blossomed in recent years, led by the "Save Indian Family Foundation" (SIFF). SIFF responds to laws and cultural norms that do not exist here—a major concern for Indian MRAs is anti-dowry laws and false claims of dowry harassment by wives. But just as in the United States' manosphere, Indian MRAs are concerned with debunking the myth of male privilege, including one advertising campaign on the Indian online magazine *Maggcom* promoting the #DontMancriminate hashtag with slogans such as "You want GENDER EQUALITY? Take it. I don't have to *hold the door*, I don't have to *hold the bags*, I don't have to *give my seat*." See Jha 2015 and Nashrulla 2015.

13. Ortiz 2015.

14. "Caucasian" is not a category on the US census, but a July 2012 estimate based on the 2010 census placed the national percentage of non-Hispanic whites at 63 percent of the overall population.

15. Valizadeh 2015f.

16. B. Smith 2016.

17. Futrelle (2017) is an excellent introduction to the MGTOW community that also addresses how the movement is implicated in racialized violence.

18. See Hart (2016) for the Alt-Right's history of itself. One of the most popular and well-known guides to the Alt-Right is the March 2016 *Breitbart* article by Allum Bokhari and Milo Yiannopoulos, "An Establishment Conservative's Guide to the Alt-Right," which I will analyze in this chapter. However, Andrew Anglin took issue with many of the analyses in that piece and countered with "A Normie's Guide to the Alt-Right" on *The Daily Stormer* in August 2016.

19. Darby (2017) is an incisive study of the few prominent women within the Alt-Right.

20. From "What Is the Deal with WMBF [white male–black female] Relationships? I Don't Get It" (Anglin 2016b); "lolwhut" is short for LOL (laughing out loud)—what? For *trigger*, see note 44 below and my discussion of the appropriative bait-and-switch later in this chapter.

21. Dewey 2014.

22. For the alt-right/alt-light distinction, see Nagle 2017, 6–9 and 11–12.

23. Valizadeh 2014c.

24. E. Hall 2008, 387.

25. One example of this narrative of the aspirational acquisition of classical knowledge can be found in Thomas Hardy's novel *Jude the Obscure*. Knowledge of the Classics was also leveraged by educated women; see Prins 2017.

26. Anglin 2016a.

27. DuBois 2001, 17.

28. Ibid., 19.

29. Yates 2015, 3 and nn.9–10. For ""Feminism Comes Full Circle," see Kavi 2015.

30. Its rise seems to have coincided with the publication of a comic strip by artist Matt Lubchansky featuring "Not-All-Man," the "lone protector of the protected" who stands vigilantly on watch to fight "reverse sexism." The history of the not-all-men defense actually goes back at least as far as 1980, with the novella *On Strike against God: A Lesbian Love Story* by feminist critic and science-fiction writer Joanna Russ.

31. All translations are mine unless otherwise noted.

32. Andramoiennepe 2016.

33. I will return to the topic of marriage laws and customs in classical antiquity in Chapter 4.

34. This effacement of women's identities is common in other Socratic texts, including the end of *Phaedo*. In the funeral oration given by Pericles in book 2 of Thucydides, he famously says that the greatest honor a woman can be given is for her name never to be mentioned in public (2.45.2). Socrates' own wife, Xanthippe, is mentioned occasionally by name but is also frequently represented as a nag.

35. A writer on *Return of Kings* using the handle "Blair Naso" did a two-part analysis of the text in 2014, titled "Xenophon's 'The Economist' Holds Valuable Lessons on a Woman's Education." He justifies the modern relevance of ancient texts in this manner: "Let's see what the wisdom of the ancients have [*sic*] to tell us. Granted, being archaic does not necessarily yield truth, but it's worth considering nonetheless." What follows is a summary of Xenophon's text with occasional offhand comments about the wretched state of femininity in the twenty-first century.

36. On gender relations in this text, see Murnaghan 1988.

37. For further scholarship on this poem, see Braund 1992 and Johnson 1996.

38. PlainEminem 2015. *Hamstering* is a reference to what is known in the Red Pill community as "the rationalization hamster," the supposed ability of women to rationalize away their indefensible behavior. "The rationalization hamster" becomes women's core defense against cognitive dissonance. Here is a definition from 2011, from the post "The Rationalization Hamster Is Now Immortal" on the manosphere blog *The Private Man*: "When a woman makes a bad decision, the hamster spins in its wheel (the woman's thinking) and creates some type of acceptable reasons for that bad decision. The crazier the decision, the faster the hamster must spin in order to successfully rationalize away the insanity."

The term *hypergamy* refers to the supposed universal female tendency to abandon her current partner if a higher-value male shows interest.

39. Chubbs 2014. He is not entirely correct, since the ancient Romans had access to effective oral contraceptives; see Riddle (1992), esp. 16–30, and Riddle (1999) on the history of oral contraception more generally. For the possibility that many women had to rear children on their own (especially during wartime), see, e.g., Rosenstein 2004.

40. For incipient programs in "male studies"—distinct from "men's studies," essentially a subset of gender studies—see Epstein 2010 and McGrath 2011.

41. This progressive view is exemplified by protests at Reed College that began in 2016. A group called "Reedies Against Racism" (RAR) protested the core humanities course, a requirement for all students, and claimed that its Eurocentrism was a form of white supremacy akin to the murders of black Americans by police officers. The course is partially, but not exclusively, dedicated to classical texts.

42. Richlin 2014, 134.

43. K. Johnson, Lynch, Monroe, and Wang 2015. The word *trigger* is colloquial shorthand for what the American Psychiatric Association's *Diagnostic and Statistical Manual of Mental Disorders* details as "internal or external cues that symbolize or resemble an aspect of [a] traumatic event(s)" that cause either "marked physiological reactions" or "intense or prolonged psychological distress at exposure"; such traumatic events include "actual or threatened

death, serious injury, or sexual violence." *Trigger* is also used as a verb for the process by which such cues set off such physiological reactions or psychological distress (American Psychiatric Association 2013).

44. Trigger warnings, also known as content warnings, are an increasingly widely discussed pedagogical tool for handling disturbing and problematic material in the classroom. The theory behind these warnings is that students who have a history of trauma—usually resulting from an experience with violence or abuse—may experience symptoms of post-traumatic stress disorder if they encounter a text that reminds them viscerally of their traumatic experience. A warning provided in advance can help the student prepare themselves emotionally or, if necessary, make the choice to refrain from engaging with that text. The efficacy and appropriateness of using such warnings in the classroom is a hotly debated topic in the academy. The American Association of University Professors' 2014 report "On Trigger Warnings" reads, "The presumption that students need to be protected rather than challenged in a classroom is at once infantilizing and anti-intellectual." Lukianoff and Haidt (2015) furthers that argument. In August 2016, Dean John Ellison of the University of Chicago sent a letter to the entering freshman class proudly announcing that "Our commitment to academic freedom means that we do not support so-called 'trigger warnings,' we do not cancel invited speakers because their topics might prove controversial, and we do not condone the creation of intellectual 'safe spaces' where individuals can retreat from ideas and perspectives at odds with their own." The letter found much support, but also strenuous criticism, including a letter from University of Chicago faculty.

45. See, e.g., Noonan 2015, Allen 2015, and Timberg 2015. Much of this criticism had blatantly sexist undertones, such as Charlotte Allen's critique on the "conservative feminist" website Independent Women's Forum: "those myths about Persephone and Daphne are now deemed too rapey [*sic*] for the delicate female flowers planted in Columbia's core 'Literature Humanities' course—so let's not read Ovid's *Metamorphoses* anymore. How about a soothing ramble through *Eat, Pray, Love* instead?"

46. None-Of-You-Are-Real 2016 and AntonioOfVenice 2016.

47. Hayward 2015.

48. Case 1985, 327. Richlin (2014) also argues that the feminist reader of a patriarchal text has three options: to ignore it, co-opt it, or replace it, and at

times she seems to come close to endorsing the third option (137). Richlin's work is part of an excellent set of scholarly articles dedicated to teaching Ovid's text with sensitivity; other articles and books to consult on this topic include Kahn (2005), Gloyn (2013), James (2014b), and Thakur (2014), along with Liveley (2012) and James (2012) on rape in Roman elegy.

49. Curtius 2015, 143–144.

50. Valizadeh 2014c.

51. Bloom 1987, 65.

52. This classification has itself been controversial: Galinsky (1992) points out, "The current catchphrase about 'dead white European males' is as inaccurate as it is racist and sexist: with reference to Greece and Rome, it is more precise to speak about light- to dark-brown Near Eastern and Mediterranean people" (116).

53. Knox 1992.

54. Alexander 2010.

55. Beale 2016.

56. Esmay 2016.

57. Dreher 2016.

58. Swann 2016.

59. Connolly (2016) is a brilliant Freudian reading of the appeal of Donald Trump; he writes, "Trump's appeal to the populace lies not in rationality but in a desire to be subjected—a masochistic attachment to an arbitrary, narcissistic, sovereign father."

2 The Angriest Stoics

1. Valizadeh 2011e.

2. Valizadeh 2015e.

3. Irvine 2008, Pigliucci 2015. Some trace the rise of interest in Stoicism back to Tom Wolfe's 1998 novel *A Man in Full*, in which a former warehouse worker, Conrad Hensley, reads the works of Epictetus while he is incarcerated.

4. As I will address later, Pigliucci has spoken about the perceived gender imbalance within the new Stoic movement, but not about the resurgence of interest in Stoicism in specifically antifeminist groups.

5. Some might find it ironic that the manosphere focuses exclusively on such a small portion of a philosophy emphasizing that everything in the universe is connected by the *logos*, a single governing principle of rationality that ensures ethics, physics, and logics are inextricable. Rosenmeyer (1986) writes on the concept of *sympatheia*: "As a consistent Stoic, one must believe that the affliction of the part affects the whole. If a man suffers, the heavens must suffer also. The fluidum of Stoic corporeality demands that the causal chains stretch in both and all directions. All constituents of the universe, without exception, share in the synergism exacted by the continuous materiality of all things. They also share in the inevitability of suffering together, of echoing and sustaining one another's injuries" (97).

6. The only figure from the Middle Stoa who receives any mention is Panaetius, because Cicero's most substantial Stoic work, *De Officiis* (*On Duties*), is a translation and adaptation of Panaetius's work about appropriate actions (*kathēkonta*, also translatable as "duties"). There is little discussion on manosphere sites of other major figures of the Middle Stoa period, such as Posidonius, or major intellectual trends, such as the increased merging with Platonic principles; see Sedley 2003, 20–24.

7. Schofield 1991; as far as concerns gender conventions, see esp. 43.

8. Cicero, in particular, is highly critical of philosophical arguments that go against entrenched social norms; see Gill 1988, esp. 193–194.

9. Sandbach 1989, 53–59.

10. Unlike Aristotelian logic, which is based on syllogisms—for example, "women are emotional, being emotional is a sign of inferiority, therefore women are inferior"—Stoic logic is propositional. So while Aristotelian syllogisms are based on universals—women, emotions, badness—the elements of the Stoic syllogism can be replaced with propositions, or "assertibles," a special subcategory of "sayables." For instance, "if a feminist writes an article, then the men of the manosphere will write abusive comments; a feminist wrote an article, so the men of the manosphere will write abusive comments." Obviously, the logic can get far, far more complex than those simplistic examples; see Bobzien 1996, 1997, and 1999. Chrysippus, the third scholarch of the Stoic school, is particularly famous for his works on logic, which underpin all Stoic thinking.

11. The Stoics believed that a divine will—God, or Providence, or Fate, or intelligent creative fire, or breath (*pneuma*)—structures all matter according to its intent. The universe is therefore highly deterministic: each thing that occurs happens because Providence intended it to be so. Anything that appears to occur randomly in fact happened because it was acted upon by a cause that we might not be able to detect. Therefore, everything in the *cosmos* is arranged in sympathy, everything acting on each other and being acted upon, all in accordance with the plan of Providence. Stoic physics is far more metaphysical, even religious, than modern-day physics. The Stoics also explicitly rejected Epicurean physics, an important concept of which is that everything is made out of atoms that move chaotically. They instead believed that matter was held together by tension (*tonos*). Early Stoics before Panaetius believed that the divine fire that created the universe would periodically purge and re-create it, an event sometimes called "the conflagration"; for primary sources on the conflagration, see Long and Sedley 1987, 46. Although Stoic logic is still an important influence on formal logic today, Stoic physics has not had much of an influence on modern-day physics. Its interest is purely historical. Fuller treatments of the topic can be found at Samburt 1987 and White 2003.

12. For primary sources on the doctrine of indifferents, see Long and Sedley 1987, 58. For a fuller description of the doctrine, see Schofield 2003, 239–246.

13. These subjects disappear from all extant sources aside from Seneca's *Natural Questions*; for the study of logic in this period, see Barnes 1997.

14. Gill 2007. He also vacillates between preferring Stoic physics and Epicurean, atom-based physics (see, for example, 4.3, 10.6, 11.18, 12.14), and he is agnostic about the conflagration (10.7).

15. Brouwer 2014, 92–135.

16. It is impossible to know whether Cicero's critique stemmed from genuine reservations of any kind or was his attempt to ventriloquize other critics of Stoicism in order to give himself a chance to rebut them.

17. On the death of Seneca, see Ker 2013.

18. R. Hughes (2011) calls him "a hypocrite almost without equal in the ancient world" (104). Most recently, Beard (2014) and Kolbert (2015) took up the same charge in their respective reviews of Romm (2014) and Wilson (2014).

19. Wilson (2014) points out that Seneca was, in theory, indifferent to exile, as he shows in his consolatory letter to his mother Helvia; however, in practice, he seems to have felt differently (77–91). Romm (2014, 28–29) has a similar thesis.

20. All Musonius translations by King (Musonius 2010).

21. Weaver 1994.

22. In spite of generally being considered a Stoic thinker, Marcus Aurelius actually only mentions Stoicism once, never mentions Zeno or Cleanthes, and mentions Chrysippus only twice (*Meditations* 5.10, 6.42, and 7.19).

23. "Mailbag: June 2015" 2015.

24. Ceporina 2012, 45–46.

25. Hadot 1995, 85. Seneca writes in one of his letters, "We need to reflect upon every possibility and to fortify ourselves against whatever hardships may come about. Run through (*meditare*) them in your mind: exile, torture, war, shipwreck" (91.7–8). The principle behind the *premeditatio malorum* practice will sound familiar to anyone who has practiced cognitive behavioral therapy; see Robertson 2010, esp. 207–226.

26. "Mailbag: June 2015" 2015.

27. AsianAway 2015.

28. Valizadeh 2015c.

29. Valizadeh 2016a.

30. Jansen 2015.

31. Black Label Logic 2016.

32. Cleary 2016.

33. Goldhill 2016.

34. Holiday 2016b.

35. Holiday has maintained a precarious balance between the mainstream media and the Red Pill community by speaking out against the ideas of controversial figures such as Donald Trump and Milo Yiannopoulos while simultaneously praising them as brilliant media manipulators.

36. "10 Great Books for Men—Volume 1" 2017. Tucker Max has an uneasy position in the manosphere. The manosphere writer who goes by the name Tuthmosis wrote on his personal blog that "Tucker Max was *the* founder of the 21st-century internet-driven masculine renaissance—a movement that we now call The Manosphere, or The Red Pill. His writing and his forum

created the first synthesis of traditional masculinity, the seduction community, and opposition to political correctness" (Tuthmosis 2014; original emphasis). However, Max has now largely fallen out of favor and has been accused of plagiarizing from Red Pill writer Mike Cernovich. The rift seems to have occurred in 2013 with the launch of Max's site *The Mating Grounds*, on which he accuses the pickup artists who worshipped him of being "sociopathic, bullshit scammers" (Max 2013). Crucially, at the time, Tuthmosis excluded Holiday from being implicated in Max's downfall.

37. Although President Trump is generally popular in the community, his occasionally interventionist foreign policy, including his bombing of Syria and Afghanistan in early 2017, is widely criticized throughout the Alt-Right, which believes he ought to focus solely on the United States.

38. Holiday 2014, 8.

39. Ibid., 46, with original emphasis.

40. "Is MGTOW the Idea of Ancient Stoicism Repeating Itself?" 2015.

41. It is unclear whether the Antipater who wrote a treatise on marriage is the Antipater who was scholarch from c. 150–130 BCE or Antipater of Tyre (first century BCE).

42. Nussbaum (2002) points out that while in literal terms the punishment was the same for both seduction (*moicheuein*) and being seduced (*moicheuesthai*), what counted as seduction was more limited for men than what counted as being seduced for women (287–288, 304–306).

43. Strategos_autokrator 2015.

44. Curtius 2015, 143.

45. Curtius 2016a.

46. Curtius 2015, 16, emphasis mine.

47. Holiday 2014; quotation on page 3.

48. Goldhill 2016.

49. Hill 2001, 26. Hill is especially hard on those who argue that Seneca is feminist: "Seneca's feminist tendencies, in particular, seem to me to be vastly overrated" (23). For those who have argued for a feminist Seneca, see Hill 2001, 23n63.

50. As he was by the Marxist historian G. E. M. de Ste. Croix (1981, 110).

51. Forney 2013.

52. Forney (2013) claims that "if girls are like gold coins, sending them to college is like dunking them in nitric acid." Since he uses this metaphor in the context of a section with the subheading "Going to College Makes Girls Less Attractive," it seems that he believes that nitric acid dissolves gold. In reality, nitric acid has no effect on pure gold and is sometimes used as a jewelry cleaner.

53. Nussbaum 2002, 303.

54. Ibid., 300.

55. Asmis (1996) makes her argument based on Antipater's analogy that a husband and wife are like two hands or two feet (76–80). However, see *contra* Engel (2003), arguing that "Antipater's ethical / political thinking leaves standard marriage practices, and parent-daughter and fiancé-fiancée relations, unchallenged and untouched" (284).

56. On the general lack of discourse in ancient literature about women sleeping with their slaves, see Parker 2007.

57. Savage 2016.

58. Asmis 1996, 71, 87ff.

59. Graver (1998) argues, citing the *Tusculan Disputations*, that *virtus* "retains its culturally determined associations with the male gender and norms of masculine behavior even while it invokes centuries of philosophical and especially Stoic thought on the ethical goal" (607–608). On *virtus* in Rome see McDonnell (2006), although Kaster (2007) articulates serious concerns that McDonnell's exclusive focus on *virtus* as a term for martial prowess is too narrow.

60. Manning 1973, 171.

61. See Graver 1998, 611.

62. Nussbaum 2002, 288–293.

63. Stoicism does, perhaps, have a better claim to be proto-feminist than any other ancient philosophy. Its only competition is Plato's critique of marriage as a social institution, in book 5 of the *Republic*, which has also been read as an early feminist text—although Stobaeus reports that Epictetus was dismissive both of it and of the women of Rome who believed that Platonic philosophy could improve their social status. Epictetus is even reported to have said that Plato "first abolishes that kind of marriage, and introduces another

kind, to the state, in its place" (Stobaeus, *Anthology* 6.58 = Epictetus fr. 15). See Aikin and McGill-Rutherford 2014, 16–17. On feminism and the *Republic*, see Pomeroy 1974, Annas 1996, and Vlastos 1997. Nevertheless, although women and men are both allowed to play the same social roles in the *Republic*, Socrates asks his interlocutors, "Do you know of any task at which the male gender does not excel above the female?" (*Republic* 455c; οἶσθά τι οὖν ὑπὸ ἀνθρώπων μελετώμενον, ἐν ᾧ οὐ πάντα ταῦτα τὸ τῶν ἀνδρῶν γένος διαφερόντως ἔχει ἢ τὸ τῶν γυναικῶν). As expected, they agree that no such task exists: although some women may be better than some men at a given task, in general, men surpass women (455d). Regardless, Plato's attitude toward women is undoubtedly more generous than that of Aristotle, who famously asserts in his *Politics* that "the male is by nature superior, and the female inferior; and the one rules, and the other is ruled; this principle, of necessity, extends to all mankind" (*Politics* 1254b). On the biological aspects of female inferiority, see Yates 2015. Sedley (2003) notes that the early Stoics had a remarkable lack of engagement with Aristotle's ideas, even though the two were roughly contemporaneous (12).

64. Cleary 2016.

65. Tomassi 2013.

66. Kimmel 2013.

67. Ibid., 2.

68. Tomassi 2014.

69. Tomassi 2014 and 2017.

70. For more on anger in the ancient world, see the essays in Braund and Most 2007.

71. "The Myth of Female Rationality" 2016.

72. See Walley-Jean (2009) for an analysis of the angry-black-woman trope and an argument that black women are, by sociological standards, *less* angry than other demographic groups.

73. Walley-Jean 2009, 83–84 and Ashley 2014.

74. Holiday 2014, 20.

75. Coates 2015, §8.

76. Jansen 2015. In Jansen's defense, Epictetus himself makes a similar point in *Discourses*, praising Diogenes for his appropriately irreverent behavior toward his master (4.1.114–117). However, Aulus Gellius tells a quite

different story in *Noctes Atticae* about one of Plutarch's slaves who attempted to reason his way out of a beating by arguing that anger was unbefitting a philosopher—to which Plutarch replied that he was not at all angry and continued the beating (1.26).

77. Engel 2003, 286–287.

78. Hill 2001, esp. 15–17.

79. Nussbaum 2002, 302.

80. Nussbaum has worked to create an improved version of the philosophy, which she calls neo-Stoicism, that would incorporate Stoicism's theorization of emotions as judgments of how much value the object of our emotion holds for us while rejecting the pure Stoic idea that such judgments should be discarded as irrational.

81. Aikin and McGill-Rutherford 2014, 10, following Hill 2001.

82. Nussbaum 1994, 322.

83. hooks 1995, 26.

84. Lorde 1981.

85. Engel 2003, 288. The Stoics are very inconsistent on reproductive rights. According to Dickison (1973), the early Stoics thought life began at birth, so abortion was morally defensible (165). However, Seneca praises Helvia for never having one, and Musonius calls abortion a crime against the gods and the state (Seneca, *Ad Helviam* 16.4; Musonius, lecture 15.30).

86. De Beauvoir 1948, 29.

87. As Vogt (2006) argues, Posidonius and Seneca both conceptualize anger as an emotional impulse that moves the individual toward revenge.

88. Jansen 2015.

89. Curtius 2015, 68.

90. Valizadeh 2016a. Valizadeh's reading of Marcus Aurelius's philosophy does not necessarily match Marcus Aurelius's history: one commenter on the post even responded, "Aurelius was far far from pacifist!" (Antonius 2016). The identification of Stoicism as proto-Christian is significant because religion and atheism are fraught topics in the community. Although many members of Red Pill communities identify as atheist, there is also a strong Christian presence in the community: one columnist on *Return of Kings*, who goes by the pseudonym Aurelius Moner, is their "resident monk." Stoicism is capable of accommodating both views: Stoic writers talk about a god,

or Zeus, who could be assimilated to the Christian god, but that Zeus also seems to be a sort of generalized divine power—the *logos*, sometimes translated "Providence," the governing principle of reason that guides the universe; see Asmis 1982, 459. On atheism and modern Stoicism, see Sulprizio 2015. On atheism in the ancient world, see Whitmarsh 2015.

91. Rutz and Rihmer 2007; Bilsker and White 2011. Some scholars believe that discrepancy is due not to more men than women being suicidal but to men being more likely than women to succeed in their attempts. See, e.g., Denning, Conwell, King, and Cox 2000.

92. Cooper 1989.

3 The Ovid Method

1. Hejduk 2014, 3, emphasis mine. All translations of the *Ars Amatoria* in this chapter are from Ovid 2014, translated by Hejduk.

2. Scholars variously date the publication of *Ars* books 1–2 between 1 BCE and 2 CE, with *Ars* 3 coming a few years later and the *Remedia Amoris* after that.

3. Hejduk 2014, 3.

4. Strauss 2005, 38.

5. "The History of Pickup and Seduction, Part I" 2016 and Weidmann 2013.

6. Ovid even wrote two poems (*Amores* 2.13 and 2.14) about his lover Corinna getting an abortion without telling him first (which initially annoys him more than the act itself). See Gamel 1989.

7. James (2003) writes, "A woman of this description—educated, intelligent, elegant, charming, independent, sexually active independent of marriage, and perpetually demanding expensive gifts—can reliably be accounted for, given Rome's class structures, only as a member of the courtesan class." (37; see also 36–41). However, there is no universal consensus on this subject, and Gibson (1998) argues that Ovid deliberately blurs the categories of *meretrix* and respectable *matrona* in *Ars* 3.

8. Myerowitz-Levine (2007) writes, "While Ovid writes for his specific Roman audience and invokes the alterity of people of other times and places, his underlying grammar is a universal one. Generations of readers who with no specialist knowledge have found very little difficulty in performing a vir-

tually automatic translation of Ovid's erotic instruction from its very Roman context into their own experience have known this all along" (258).

9. Volk 2007, 236.

10. Burns 2014; Strauss 2005.

11. This view is informed by evolutionary psychology. Strauss (2005) writes that "among the required reading for all PUAs were books on evolutionary theory: *The Red Queen* by Matt Ridley, *The Selfish Gene* by Richard Dawkins, *Sperm Wars* by Robin Baker. You read them, and you understand why women tend to like jerks, why men want so many sexual partners, and why so many people cheat on their spouses" (294).

12. Volk (2007) gives the most succinct argument against this viewpoint: "Being a social practice, the love taught in the *Ars Amatoria* is not something universal, experienced equally by all people everywhere, but a phenomenon that is shaped by place, time, and circumstances. In other words, it is a cultural construct: any culture, any society will define and practise 'love' differently" (242).

13. Burns 2015.

14. Ainsworth 2015.

15. Ibid.

16. Kahn 2005. Kahn describes her surprise when a student labeled the text in this manner: "With her question, the student was suggesting that rape was what the book was primarily about, and that far from being critical of the rapes he described, Ovid was offering instructions on how to imitate them. This was a new perspective on what has traditionally been regarded as a wry compendium of stories about the changeableness of human nature and the fickleness of the gods" (1).

17. Thorn 2012, 21–22.

18. Ibid., 24.

19. Krauser n.d., with original emphasis.

20. The *praeceptor* goes on to acknowledge that those wings did not serve Icarus very well, without recognizing what the analogy says about his own methodology—one of many hints that he is not quite as all-knowing as he would like to appear, and that *ars* may have its limits. On this episode, see Myerowitz 1985, 151–174 and Sharrock 1994, 87–195.

21. Volk (2007) agrees: "The *Ars Amatoria* is thus really something like the art of dating, the art of the love affair, and *not* the art of love. This, of course, is the reason why it is teachable in the first place: *amor* is for Ovid not a feeling but a mode of behaviour, and thus can be mastered by following the simple steps laid out in his didactic poem" (242).

22. Specifically, the *amator* is a stock character in Roman comedy, a young man whose defining feature in the play is his sexual pursuit of young woman (either a *meretrix* or a *virgo*).

23. Although even von Markovik's emotional abuse is considered harmless compared to the legendary October Man sequence, a PUA technique that uses hypnosis to compel women to sleep with men regardless of consent.

24. Arrowsmith 2014, 73.

25. von Markovik 2007, 96–97.

26. Fleishman 2013.

27. von Markovik 2007, 50.

28. Strauss 2005, 142.

29. Ibid., 38; Jeffries 1992, 125. This provides a built-in escape clause if the reader does not succeed: the author can just claim he did not try hard enough. Von Markovik (2007) recommends that newbies open a staggering two hundred sets (that is, introduce themselves to two hundred groups of girls) in a month to get acclimatized; if the student opens fewer sets than the suggested number and fails to become a successful pickup artist, the methodology is not to blame (40).

30. Strauss (2005) later goes into even more effusive detail about Jeffries's repugnant physical appearance: "He was thin and gawky with gray stubble and greasy skin. His hairline was a receding mop of short, unkempt, ash-colored curls, and the hook in his nose was so pronounced he could have hung his overcoat on it" (45).

31. Strauss 2005, 8.

32. Although the last element in his full name, Publius Ovidius Naso, suggests that, as with Jeffries and Strauss, Ovid's nose may not have been his best facial feature (*naso* means "nose"). On the Roman adoption of unflattering nicknames as *cognomina*, see Salway 1994, esp. 127. Ovid's cognomen was inherited, so it is possible that he inherited the name but escaped the ugly nose;

although he refers to himself as "Naso" quite frequently, he likely does so because it fits the meter.

33. S. Kendall (2014) addresses these issues, advising Asian men to "transcend your race."

34. Donovan Sharpe writes about seduction from a black male perspective on *Return of Kings*, with articles such as "The Pros and Cons of Game for Black Men in the West" (2014) and "5 Tips for Non-White Men Who Want to Date White Women" (2016).

35. The truly hopeless are even called *omega males*, although Theodore "Vox Day" Beale prefers the term *gamma male*.

36. Valizadeh n.d.

37. Watson 2007.

38. Strauss 2005.

39. Valizadeh 2011b, 68.

40. Valizadeh 2012b.

41. Valizadeh 2011d.

42. Ibid.

43. Galbi 2010b.

44. Galbi 2010a.

45. Galbi 2010b.

46. Galbi 2010a.

47. Gaisser (2002) writes that ancient texts "are not teflon-covered baseballs hurtling through time and gazed up at uncomprehendingly by the natives of various times and places, until they reach *our* enlightened grasp; rather, they are pliable and sticky artifacts gripped, molded, and stamped with new meanings by every generation of readers, and they come to us irreversibly altered by their experience" (387, original emphasis).

48. Nietzsche 1990, §52.

49. Galbi 2010b.

50. Peter Burns, the writer of "Lessons from PUA Ovid" did a longer and more extensive comparison on his own blog, *Renaissance Man Journal* (2014).

51. Although von Markovik (2007) is somewhat unique among pickup artists in that he favors what he calls "peacocking," which he defines in his glossary as "wearing outlandish clothes in the field to advertise your survival

abilities to women. Being unaffected by the social pressure that this creates demonstrates higher value." However, he cautions, "'peacocking' should be a unique and dominant expression of higher social value; otherwise, it not only loses its desired effect but also can actually work against you in the field. One guy dressed to the nines looks like 'the man'; two guys decked out that way look gay" (26). Most photographs of von Markovik show him wearing eyeliner and either goggles or a top hat made of colored fur.

52. See Walters (1997) on the impenetrability that was legally afforded to Roman men—but, crucially, not to women.

53. See, for example, von Markovik 2007, 136–142.

54. Weidmann n.d.

55. Hoinsky 2013.

56. Bustillos 2013.

57. Hoinsky 2013.

58. Strauss 2005, 102.

59. von Markovik 2007, 104.

60. Fahrenthold 2016.

61. In Valizadeh's post "5 Reasons Your Game Sucks," the third is "you use alcohol as a crutch" (2011a). Ovid also mentions the hazards of alcohol, saying, "Here, don't put too much confidence in the treacherous lamp:/night and pure wine wreak havoc on judgments of beauty" (*Ars* 1.245–246).

62. Lyaeus is another name for Dionysus, the god of wine, here used metonymically to refer to wine itself.

63. Sharrock (2007) notes that in the *Ars Amatoria*, "Women are supposedly being taught how to catch and keep (and exploit) a lover, but it is hard to resist the feeling that what they are actually being taught is how to let their men have it all ways" (28).

64. Valizadeh 2011d.

65. See, for instance, von Markovik (2007): "It is easier to attract a new woman than to fix things when they go wrong with your existing target" (39).

66. Weidmann n.d.

67. Even though the *Remedia* positions itself as the opposite of the *Ars*, it is more of a continuation or sequel than a reversal; see Rosati 2007, 151–157. The two texts even offer some of the same advice: consider keeping a harem in-

stead of being monogamous, if for no other reason than to make your target jealous (*Ars* 2.425–460; *Remedia* 439–488).

68. Weidmann n.d.

69. Hexter 2007.

70. Valizadeh 2014b.

71. It may seem absurd to equate emotional abuse with cosmetics, but the comparison is actually a common trope in responses to critics of pickup artistry. When Valizadeh was denied entrance to Australia in 2016, a writer using the pseudonym Thomas Hobbes published a satirical gender-bent response, "International Outrage: Men around the World Try to Stop Make-Up Artist Lecture Tour," on *Return of Kings* (Hobbes 2016). Ovid wrote an entire treatise, the *Medicamina Faciei Femineae* (*Cosmetics for the Female Face*), giving makeup advice to women; M. Johnson (2016) is an excellent recent study of the text.

72. The *lex Iulia de adulteriis coercendis* also permitted fathers and husbands to execute adulterous wives and their lovers under some circumstances. On Augustus's moral legislation, see Mette-Dittmann 1991.

73. Barchiesi 1997, 4.

74. Gibson 1999.

75. Barchiesi 2007. Furthermore, despite his moralizing legislation, Augustus was reputed to be a serial adulterer himself; Suetonius shares a few choice stories in his *Life of Augustus* (69–71).

76. Welch (2005) analyzes the use of *amor* as a palindrome for *Roma* in Propertius; the same principle could be operating here for Ovid.

77. Valizadeh (2013) argues in favor of expatriation.

78. Weidmann 2016a and 2016d.

79. Preston (2010) writes that "nearly all of the PUAs in 'The Game' are white and the 'seduction community' could be broadly described as a micropolitical movement for white genetic survival" (338).

80. Galbi 2010b. Ali was a prolific blogger for a long time, but has recently turned to writing for larger manosphere websites in articles such as "Why I Became a Black Men's Rights Activist" (2015b) and "5 Reasons Why Black Feminism Is a Failure" (2015a).

81. See, e.g., Sharpe 2014 and 2016.

82. Frantzen 2016.

83. Ibid.

84. Strauss 2005, 214.

85. "If a girl is with a boy, assume they are just friends and approach the group, initiating the chat with him. Befriend him. Once you've disarmed him, you've reached the waypoint where you ask, 'How do you know each other?' He will tell you. If he's the boyfriend, you just made a new bud and didn't even introduce yourself to the girl, so you can't get in trouble. If he's not the boyfriend, she is now fair game" (von Markovik 2007, 119).

86. Strauss 2005, 423.

87. There has been a great deal of scholarship on this topic, but the foundational texts remain Foucault (1978–86); Halperin, Winkler, and Zeitlin, eds. (1990); and Davidson (2007) on Greek sexuality; and Hallett and Skinner (1997) on Roman sexuality. More recently, see Hubbard 2014.

88. The role and manipulation of gender may be the single most-studied aspect of Latin elegy, and there is a mountain of scholarship on the topic. A good introduction is Greene (2012); see also the essays in Ancona and Greene 2005. Wyke (2007) puts it well: "Given that Roman sexual relations were constituted in terms of activity and passivity, of domination and subordination, of superiority and inferiority, engendered as masculine and feminine, and aligned with the relationships of master and slave, the persistent Propertian strategy of casting the male lover in a submissive, servile role in relation to his obdurate mistress disturbs the gendered protocols of Roman sexuality. The male *ego* enacts the role of a faithful, submissive, subordinate woman" (168).

89. For Catullus's troubled position on the border of the canon of Roman elegists, see Wray 2012.

90. On masculinity in Catullus, see Wray 2001, esp. 64–112.

91. Strauss 2005, 9.

92. Greene (2000) addresses this positioning of the relationship between Propertius and Maecenas as one of *amor* (250–251).

93. Fear 2000, 220.

94. "This is what the *Ars* does to us: it holds on to us, controls us, keeps us spellbound, embraces us with the erotics of its text" (Sharrock 1994, 24).

95. Strauss 2005, 41.

96. Ibid., 87.

97. Ibid., 398.

98. Strauss 2015.

99. Ali 2013.

100. Wyke 1987.

101. Valizadeh 2012a, 71.

102. This interpretation is suggested by Liveley (2012), although less as an excuse for Ovid than as a pedagogical technique to help students in the college classroom approach Ovid in a non-triggering way (545–546). James (2012) also argues that Ovid is "exposing predatory male sexuality" rather than advocating it (554).

103. This idea is perhaps best expressed by the user "redpillschool" on Reddit—later identified as former New Hampshire politician Robert Fisher—who wrote, "Every woman wants to be attractive enough to be raped. It's like the pinnacle of male desire, when he can't stop no matter what" (R. Fisher 2015).

104. Valizadeh 2010.

105. Ibid.

106. Denes 2011.

107. Valizadeh 2012a, 28.

108. Valizadeh even argued in the infamous blog post "How to Stop Rape" that legalizing rape would give women an incentive to control their alcohol consumption (2015b).

109. Valizadeh 2015a.

110. Thorn 2012, 45–46.

111. Zadrozny 2016.

112. A. Smith 2012.

113. Zadrozny 2016.

114. Strauss 2005, 241; von Markovik 2007, 2. The true English adjective derived from the name of the goddess is *venereal*, but it is understandable that he chose not to call his methodology the venereal arts, even if it would have been more accurate.

115. Cahoon (1988) argues that Ovid favors this metaphor because he conceptualizes love as being violent. Sharrock (2007) likewise points out that all the mythological digressions in *Ars* 1 engage in "romanticization of force" (32).

116. Their reasoning, of course, is pseudo-evolutionary: women are disincentivized from engaging in sexual activity because of the possibility of pregnancy.

117. Jeffries 1992, 2 (original emphasis), 4.

118. In addition to the original post suggesting this opener, Weidmann wrote a followup post with screenshots a reader sent proving the technique works (Weidmann 2016c and 2016b).

119. Sexual satisfaction seems unlikely, considering that Valizadeh disseminates advice such as "it doesn't matter if she orgasms or not" (Valizadeh 2008). Ovid advises women that, while they *should* enjoy sex as much as men, if they don't they should fake it as convincingly as possible (*Ars* 3.797–804).

120. Analyzed brilliantly in Baker 2013.

121. Valizadeh 2011c, 4–5.

122. Ibid., 6.

123. Ibid.

124. Ibid., 10.

125. Ibid., 20.

126. Ibid., 38.

127. Ibid., 64.

128. Weidmann 2017a.

129. Valizadeh 2011d.

130. Weidmann 2013.

4   How to Save Western Civilization

1. Leonid 2017.

2. Valizadeh 2015d.

3. Valizadeh 2014a.

4. Elam 2010, original emphasis.

5. Supremo 2016, original emphasis.

6. A few years ago I also inadvertently created experimental conditions whose outcomes strongly suggest that the manosphere does not want to talk about Phaedra. In July 2015, I published the article "He Said, She Said: The Mythical History of the False Rape Allegation" on the feminist website *Jezebel*. That article quickly passed 100,000 views. I later learned that some of those

views likely came from a post on the subreddit r/mensrights, a community with over 170,000 subscribers, titled "Hell Has Frozen Over: Jezebel on False Rape Accusations," implying that my article was an anomalous recognition by a feminist website of the fact that false rape allegations exist (Eaton80 2015). My use of ancient literature was entirely ignored; one commenter even wrote, "I stopped reading halfway through as she rambled on about phaedra. which I'm sure was jizzabel's (sic) intent so they can say 'hey, we acknowledged that false rape accusations happen, see we wrote a post about it.'" Another complained that the piece was "littered with so many unnecessary references to fiction."

7. Leonid 2017.

8. This kind of argument is particularly common in response to fictional stories about false allegations, including the feminist rejection of the movie (and, to a lesser extent, the book) *Gone Girl*. Feminists said the film was misogynistic and toxic, and that it reinforces the mistaken idea that false rape allegations are commonplace. For example, feminist Joan Smith wrote in *The Guardian* that "*Gone Girl*'s recycling of rape myths is a disgusting distortion" (Smith 2014). Unsurprisingly, the Red Pill community loved the film. On the *A Voice for Men* forum, one poster said the film was "AN ABSOLUTE MUST. Especially if you're interested at all in the red pill/MRM/MGTOW movement. . . . The most red pill movie I've seen in a very, very long time" (artlone 2014).

9. Valizadeh 2015g, my emphasis.

10. Ibid.

11. Wheatcroft and Walklate 2014, 246. Cognitive ease may also play a role in the tendency to pigeonhole accusers into specific roles: as Crenshaw (1992) argues, there is a tendency "to cast the complainant in one of several roles, including the whore, the tease, the vengeful liar, the mentally or emotionally unstable, or, in a few instances the Madonna. Once these ideologically informed character assignments are made, the 'story' tells itself, usually supplementing the woman's account of what transpired between the complainant and the accused with a fiction of villainous female intentionality that misleads and entraps the 'innocent' or unsuspecting male" (408).

12. Kanin 1994. For a critique of Kanin, see Lisak (2011), which argues, "Kanin's 1994 article on false allegations is a provocative opinion piece, but it

is not a scientific study of the issue of false reporting of rape. It certainly should never be used to assert a scientific foundation for the frequency of false allegations."

13. Although it is supported by the findings of a 2010 study; see Lisak et al. 2010. A more recent Buzzfeed article about the Baltimore County Police Department's disproportionate number of rape reports marked "unfounded" by the police, about thirty-four percent, puts the percentage of actual false allegations at seven percent (Campbell and Baker 2016). The article uses more recent FBI data.

14. Furioso 2016.

15. McArdle (2015) argues that "rape statistics are a mess" and ends with the conclusion, "Until we get mind reading machines, the only thing we can know about the actual prevalence of false rape reports is that we don't know it."

16. Furioso 2016.

17. Erdely 2014.

18. Saunders (2012) addresses the divergence between the police definition of false allegations and the research definition.

19. Wheatcroft and Walklate 2014, 240–241.

20. Herman 1992, 72.

21. In *Phaedra's Love*, a play by Sarah Kane very loosely based on Seneca's *Phaedra*, Phaedra's false accusation of rape follows a very real and emotionally abusive sexual relationship between her and Hippolytus. Kane later said that "what Hippolytus does to Phaedra is not rape—but the English language doesn't contain the words to describe the emotional decimation he inflicts. 'Rape' is the best word Phaedra can find for it, the most violent and potent, so that's the word she uses" (I. Ward 2013, 235). On *Phaedra's Love* and rape myths, see Ward 2013, 234–237.

22. Saxton 2015, emphasis mine.

23. Dworkin 2006, 158–159.

24. DiKaiomata 2015.

25. Patai and Koertge 1994, 124.

26. MacKinnon 1989, 174.

27. The role that race and class play in Red Pill ideas about sexual assault is almost entirely implicit—as, for example, in Leonid's assertion, quoted at

the beginning of this chapter, that false accusations are "a standard rite of passage for any man of worth," without defining precisely what he means by "worth."

28. They might, for example, have chosen Homer's works instead. Aristotle does distinguish between history and poetry, but not between history and myth: the difference is one of genre, not content. He writes in the *Poetics* (1451b) that history deals with "what things happened" whereas poetry deals with "the sort of things that might happen" (τὸν μὲν τὰ γενόμενα λέγειν, τὸν δὲ οἷα ἂν γένοιτο).

29. Redfield 1985, 100.

30. J. Hall (2002) analyzes Herodotus's definition of "Hellenicity" as shared blood, language, religion, and customs (189–194).

31. Walcot (1978) explores rape in Herodotus. So does Harrison (1997), although he only briefly touches on the beginning of the book; he is more interested in the scenes of what he calls "hard-core rape." Sansone (2016) lists the historians who have considered Herodotus's reciprocal abductions "humorous" or parodic (2–3).

32. Epictetus, a Stoic philosopher whose works I analyzed in Chapter 2, imagined Paris and Menelaus as close friends until "a tempting bait, a pretty girl, was thrown between them" (*Discourses* 2.22.23).

33. Euphiletus knew that the affair was taking place and arranged to catch them in the act, so the act falls on the border between a crime of passion and cold-blooded murder. For the possible fictionality of the text and its many parallels with tropes from comedy, see Porter 2007.

34. Lysias 1.32–36. Harris (1990) insists that this was merely a rhetorical position, while Omitowoju (2002) takes it far more seriously as signifying the role of female consent in Athenian sex (66–68).

35. Gorgias's "Encomium of Helen" is the best-known example of this type of speech; Euripides's *Trojan Women*, lines 914–965, is another.

36. On rape as an element of war in the ancient world, see Gaca 2014. The only insight into Briseis's subjectivity that we are given comes much later in the *Iliad*, when Briseis mourns over Patroclus's body and mentions that Patroclus had told her Achilles would marry her.

37. Although, in the preface to *Ab Urbe Condita*, Livy seems to express some doubt over the veracity of the tale about Romulus's divine parentage,

and notes that Rome has claim to a demigod founder primarily because of its vast power.

38. All analyzed in Arieti 1997.

39. Although Moses (1993) argues that, by choosing rape as the lesser of two evils, Lucretia may be seen to have "consented" under Roman law (40–43).

40. For more on the complex concept of *hybris*, see Fisher 1992.

41. Since my focus here is on comparing the sexual politics of heterosexual rape and marriage in the ancient world and today, I will not delve into the legal complexities regarding the rape of men or boys. Much has been written about the nature of pederasty in the ancient Greek world and the degree to which it was a coercive relationship: there was a mentorship aspect to these relationships—the older partner instructed the younger on how to be an excellent citizen—but they were also sexual relationships. These relationships often entailed the molestation of very young boys—barely teenagers, since the onset of facial hair was considered a sign of aging out of the *erōmenos* (beloved) role, although Ferrari (2002) provides a more nuanced view than the typical dichotomy of bearded and unbearded, arguing that the various stages in the gradual appearance of facial hair marked various phases of attractiveness (127–140). The classic works on the subject of pederasty (and on Greek homosexuality more generally) are Foucault 1985, Dover 1989, and Halperin 1990. Lear (2014) provides a very useful overview of the ancient evidence for pederasty. Davidson (2007) broadens the age range for the *erōmenos* role substantially to argue that many same-sex relationships occurred between men who were close in age, but this claim has been highly controversial.

42. Harris 2004, 60–63.

43. Including, among others, Plautus's *Aulularia* and Terence's *Adelphoe*. Another common plot of New Comedy involves a newlywed husband threatening to leave his wife when he realizes she became pregnant before they were married; the tension is dissolved when it is discovered that he was in fact her rapist and just had not yet realized it. This scenario occurs in Menander's *Epitrepontes* and Terence's *Hecyra*.

44. See Just 1989, 28–52; Sealey 1990, 25–40; and Patterson 1991.

45. Additionally, if the woman was an heiress (*epikleros*) whose father died with no sons, her nearest male relative could dissolve her current marriage and marry her in order to gain control of the estate; see Sealey 1990, 29.

46. On slaves and sexuality, see Cohen 2014.

47. *Epistulae ad Atticum* 5.4.1.

48. Saller 1987, 33. Saller estimates that, for all other classes, the numbers would be closer to two-fifths of women and one-fifth of men. However, other scholars estimate a slightly lower age at first marriage, including Scheidel (2007). The ancient legal sources are *Tituli Ulpiani* 5.2 and Ulpian de Sponsalibus, *Digest* 23.2.2. The latter states clearly, "A marriage is not valid unless everyone agrees, that is, both those who marry and those in whose power they are" (*nuptiae consistere non possunt nisi consentiant omnes, id est qui coeunt quorumque in potestate sunt*).

49. Ulpian also says that the bride may only refuse when the groom is unfit in some manner, further calling into question her freedom of choice (Ulpian de Sponsalibus, *Digest* 23.1.12). There were two kinds of marriage, *cum manu* (in which guardianship of the woman passed to her husband) and *sine manu* (in which guardianship remained with the father or *tutor*. The evidence suggests that, over time, marriage *cum manu* became increasingly uncommon. On the complexities of Roman marriage law, see Treggiari 1991.

50. Ted Hughes translating *Metamorphoses* 5.524–527 (Ovid 1997, 59).

51. Packman 1993.

52. Harris 2004. James (2014a) calls out the tendency to focus on rape in the context of "its social and cultural meaning" rather than its personal consequences: "The shock and trauma of the event, at the moment, are experienced bodily rather than culturally" (29, 33). See also Gardner 2012. Walters (1997), however, points out that penetrability and violability were crucial aspects of how the female body was imagined in ancient Rome, suggesting that the experience of being penetrated may have meant something different to ancient women than it does today.

53. Harris 1997, 483; Deacy and Pierce 1997. Harris goes on to clarify that "this statement is not meant to be facetious," but it is certainly dismissive.

54. Leonid 2017.

55. Phaedra first appears in the *Odyssey*, as one of the many shades Odysseus encounters in book 11. She is only a name—her family and history are not mentioned. *Hippolytus* was part of a trilogy that earned Euripides first prize in the tragic competition—one of only five first prizes he was awarded,

compared to the eighteen won by his competitor Sophocles and thirteen by his predecessor Aeschylus.

56. This kind of argument is still widely used to refute allegations today; in 2016, Donald Trump responded to sexual assault allegations by *People* writer Natasha Stoynoff by saying at a campaign rally in October 2016, "You take a look. Look at her. Look at her words. Tell me what you think. I don't think so. I don't think so." Such arguments center men and imply that, in the case of an allegation of rape, female desirability is directly correlated to credibility. The men of the Red Pill, who often tell women on Twitter that they are too ugly or fat to rape, find this kind of argument compelling.

57. Kohn (2008) argues that the idea of Theseus having three wishes from Poseidon was invented by Euripides for this story.

58. See *contra* Roisman (1999a) following Fitzgerald (1973) for an extensive argument that Phaedra is manipulative and deceitful. Roisman (2005) furthers this argument through a comparison with Seneca's *Phaedra* and argues that "her long refusal to tell the nurse what is bothering her does not stem, as it might seem, from shame or modesty, but is a calculated tease designed to draw the nurse into her troubles. . . . In short, Euripides presents Phaedra as a lustful and scheming woman, determined to get her man, and with the skill and control to manipulate her doting nurse into telling him of her love" (74–75).

59. Contrast, for example, the chorus of Euripides's *Medea*, who seem terrified of Medea even as they acknowledge that she has been wronged by her faithless husband.

60. The *Hippolytus* that still exists was Euripides's second version of the play, and the earlier, now-lost Phaedra was more sexually aggressive and more malicious. The evidence for the relationship between the two and for the dating of the extant *Hippolytus* comes from the play's ancient hypothesis, attributed to Aristophanes of Byzantium. Although it is by far the consensus—and see McDermott (2000) on why it should continue to be the default assumption—a few scholars have been skeptical or at least advised caution in assuming the precedence of the lost play, including Gibert (1997) and Hutchinson (2004). Roisman (1999b), while not disputing that the lost play was the earlier of the two, argues that the shamelessly forward proposition was not sexual in nature, but rather political: Phaedra suggested to

Hippolytus that they might marry and usurp the throne together. We do not know why Euripides chose to write a second tragedy telling the same story with a less objectionable Phaedra. Whatever his reasoning, the second play did not entirely absolve Phaedra of her sins; more than fifteen years after the performance of Euripides's *Hippolytus the Wreath-Bearer* (the existing version with a more innocent Phaedra), the women in Aristophanes's comedy *Thesmophoriazusae* continue to complain about Phaedra as though she were the paradigmatic evil woman.

61. The *Ars Amatoria* reference occurs in the context of advice the *praeceptor amoris* gives his students about how to cultivate their appearance. As proof that men do not need to worry excessively about grooming themselves, he notes, "Phaedra loved Hippólytus, and he wasn't well groomed" (Ovid 2014). As with most mythical examples in the *Ars* that I mention in the previous chapter, this one fails to inspire confidence that the *praeceptor*'s advice will lead to successful and happy results.

In *Metamorphoses*, Hippolytus tells the story to the nymph Egeria, sparing a single line to speculate whether Phaedra was motivated to make her false accusation by fear of discovery or spite at being rejected (*indiciine metu magis offensane repulsae?*) before spending twenty-five lines narrating his chariot accident and what it felt like to have his legs crushed (*Met.* 15.503).

62. Roisman (2005) argues that "in highlighting Theseus' philandering, Seneca leads his audience to wonder whether she might have been less drawn to her misogynistic stepson if his father had been a more faithful husband" (76).

63. Seneca's personal brand of Stoicism often falls short of the purest form of the philosophy—a deficiency that, as I discuss in Chapter 2, Seneca is not afraid to admit to. On the literariness of Senecan heroes, see Boyle 1997, 133–137. The precise amount of Stoicism in Senecan drama has been much discussed; for a bibliography, see McAuley 2012, 65nn.1–2. Armstrong (2006) argues that the play shows "Seneca's questioning of the Stoic doctrine he elsewhere accepts" (290).

64. Seneca's Phaedra also faces punishment more severe than Euripides's Phaedra did. Legally speaking, in Euripides's time Phaedra would have been guilty only of adultery, whereas the desires of Seneca's Phaedra go against Roman taboos about incest. McAuley (2012) writes that "marriage between

steprelations at Rome was outlawed under Augustan legislation, rendering any sexual alliance between stepmother and stepson, in principle at least, incest. By contrast, in fifth-century Athens, although there was fear of stepmothers' supposedly unbridled sexuality, an incest prohibition does not seem to have applied, possibly reflected in the fact that Euripides's Phaedra is never described as stepmother; instead the chief moral focus on her is as would-be adulteress" (39–40). McAuley also notes that Euripides never uses the Greek word for stepmother, *mētruia*, to describe Phaedra—making Seneca's focus on her status as Hippolytus's *noverca* all the more striking.

65. It is not until Racine's *Phèdre* that Hippolytus is given a love interest of his own, Aricia.

66. Cairns (1993) analyzes the use of the term *aidos* in the *Hippolytus* more generally (330–332).

67. For *aidos* as a pleasure, see, e.g., Segal 1970; Kovacs 1980; Craik 1993; Williams 1993, 225–230; and Furley 1996. The women of Greek tragedy often police each other's *aidos*. In Euripides's *Trojan Women* (415 BCE), Andromache boasts about how much more virtuous she was than the average wife, saying, "Whatever is considered virtuous for a woman, I toiled to do in Hector's home. First, whether a woman is criticized or not, not remaining in the house brings ill repute, so I suppressed my desire to leave and stayed inside. And I did not allow womanly gossip into my home, but instead contented myself with an honest mind" (*Trojan Women* 645–652).

68. McAuley (2012) makes a similar point while contrasting Seneca's Phaedra with Medea: "Phaedra does not overtly confront the androcentric socio-political order that constrains and categorizes her. Rather, she tries to work within it, to find some way in which it might accommodate her aberrant and excessive desires, perhaps even (as she deludes herself at one point) legitimate them through marriage (57; *Phaed.* 597)."

69. Burt 1980, 217; Brownmiller 1975.

70. C. Ward 1995, 45.

71. Valizadeh 2014a.

72. Ibid.

73. Ibid.

74. MacKinnon 1987, 5; Valenti 2014; and Young 2014.

75. Valizadeh 2014a. One particularly stunning example of this trend is the three-month jail sentence Judge Aaron Persky gave to convicted rapist and Stanford swimmer Brock Turner in a widely publicized 2016 incident. Even though two witnesses saw Turner assaulting an unconscious woman—and tackled him as he tried to escape—Persky elected to give Turner a light sentence, citing concern for his future.

76. Wheatcroft and Walklate 2014, 242, with original emphasis.

77. Valizadeh 2014a. This mechanism is constant across all of Euripides's false accusation plays, as well as in the very similar story in Genesis 39 about Potiphar's wife and Joseph. The false accuser's only power is the power to influence men.

78. Valizadeh 2015g.

79. Ibid., emphasis mine.

80. Ibid., emphasis mine.

81. Valizadeh 2017.

82. Anglin 2017, emphasis mine. The white knight is a common Red Pill trope; see the Glossary.

83. Weidmann 2017b.

84. Elliot Rodger, for instance, a twenty-two-year-old who killed six and injured fourteen in 2014 in Isla Vista, California, identified as an incel.

85. E. Hall 2015.

86. The women complain about other characters as well, including Stheneboea, the mythical queen of Argos whose sexual advances were rejected by the Pegasus-riding hero Bellerophon. As revenge, Stheneboea told her husband Proteus that Bellerophon had raped her, and Proteus set in motion a series of events meant to lead to Bellerophon's death. The story has marked similarities to Phaedra's story, although Stheneboea is a more straightforward villain. Euripides, like the men of the manosphere, seems to have had something of an obsession with false rape allegation myths. He wrote at least five tragedies about them, a small but significant fraction of his lifetime output of about ninety plays. In addition to the two *Hippolytus* plays and *Stheneboea*, the plays to which the women in *Thesmophoriazusae* vocally object, Euripides probably wrote at least two other lost plays with similar themes: *Peleus* and *Tennes*. *Peleus*, a tragedy about the father of the Trojan War hero Achilles,

likely told a story from the hero's youth about how he was falsely accused of rape by Astydameia, the wife of king Acastus, who was hosting him. The same myth is depicted by the lyric poet Pindar (*Nemean* 5.26–34). *Tennes* was about a lesser-known mythical hero who, like Hippolytus, was accused of rape by his stepmother. The story is told by the Greek travel writer Pausanias (10.14.2).

87. Sommerstein (1994) comments on these lines that "the point may be . . . that Phaedra's particular villainy (revenging herself, by a false accusation of rape, on a man who had spurned her) is not one of those that women now habitually practise." If that analysis is correct, it might suggest that false rape allegations were as infrequent in Athens as they are today, no matter how often Euripides wrote about them.

88. Weidmann 2017b.

Conclusion

1. Padilla Peralta 2015.
2. Zuckerberg 2016.
3. Grant 2016; Curtius 2016b.

REFERENCES

"10 Great Books for Men—Volume 1." 2017. *Illimitable Men*, June 27. https:// illimitablemen.com/the-library/books-for-men/.

Aikin, S. and E. McGill-Rutherford. 2014. "Stoicism, Feminism and Autonomy." *Symposion* 1: 9–22.

Ainsworth, C. 2015. "Sex Redefined." *Nature* 518: 288–291.

Alexander, M. 2010. *The New Jim Crow: Mass Incarceration in the Age of Colorblindness*. New York: The New Press.

Ali, M. 2013. "Quick Hit: Don't Let the Dime Piece Be the Enemy of the Plain Jane." *The Obsidian Files*, September 11. http://obsidianraw.bravejournal .com/entry/137558.

———. 2015a. "5 Reasons Why Black Feminism Is a Failure." *Return of Kings*, April 27. www.returnofkings.com/62306/5-reasons-why-black-feminism -is-a-failure.

———. 2015b. "Why I Became a Black Men's Rights Activist." *A Voice for Men*, January 29. www.avoiceformen.com/sexual-politics/why-i-became-a -black-mens-rights-activist/.

Allen, C. 2015. "Ovid's Metamorphoses Now Deemed Too 'Triggering' for Students at Columbia." *Independent Women's Forum*, May 13.

American Association of University Professors. 2014. "On Trigger Warnings." Report. Washington, DC: American Association of University Professors. www.aaup.org/file/2014-Trigger_Warnings.pdf.

American Psychiatric Association. 2013. "Diagnostic Criteria for Posttraumatic Stress Disorder," in *Diagnostic and Statistical Manual of Mental Disorders*, 5th ed., 309.81 (F43.10). Washington, DC: American Psychiatric Association.

Ancona, R. and E. Greene, eds. 2005. *Gendered Dynamics in Latin Love Poetry.* Baltimore: Johns Hopkins University Press.

Andramoiennepe. 2016. "To Marry or Not to Marry? An Ancient Perspective." *The Red Pill.* Reddit. www.reddit.com/r/TheRedPill/comments/4smquk /to_marry_or_not_to_marry_an_ancient_perspective/.

Anglin, A. 2016a. "A Normie's Guide to the Alt-Right." *The Daily Stormer,* August 31. https://dailystormer.name/a-normies-guide-to-the-alt-right/.

———. 2016b. "What Is the Deal with WMBF [white male–black female] Relationships? I Don't Get It." *Daily Stormer,* September 19. https:// dstormer6em3i4km.onion.link/what-is-the-deal-with-wmbf -relationships-i-dont-get-it/.

———. 2017. "White Sharia in Action: Nathan Damigo's 'Punch Heard round the World.'" *The Daily Stormer,* April 16. https://dstormer6em3i4km.onion.link /white-sharia-rising-nathan-damigos-punch-heard-round-the-world/.

Annas, J. 1996. "Plato's *Republic* and Feminism." In J. Ward 1996, 3–12.

Anton, M. (Publius Decius Mus). 2016. "The Flight 93 Election." *CRB Digital,* September 5. www.claremont.org/crb/basicpage/the-flight-93-election/.

AntonioOfVenice. 2016. "Remember When We Laughed at SJW Students Calling Ovid 'Problematic' and 'Triggering'? The University Caved: Ovid Has Been Removed from the Syllabus." *Kotaku In Action.* Reddit. www .reddit.com/r/KotakuInAction/comments/3tto1n/remember_when_we _laughed_at_sjw_students_calling/.

Antonius, M. 2016. Comment on Valizadeh 2016. http://www.rooshv.com /marcus-aurelius-meditations-is-the-best-manual-we-have-on-how-to -live#comment-2547458854.

Arieti, J. 1997. "Rape and Livy's View of Roman History." In Deacy and Pierce 1997, 209–229.

Aristophanes. 1994. *Thesmophoriazusae.* Edited by A. Sommerstein. Warminster: Aris and Phillips.

Armstrong, R. 2006. *Cretan Women: Pasiphae, Ariadne, and Phaedra in Latin Poetry.* Oxford: Oxford University Press.

Arrowsmith, A. 2014. "Rethinking Misogyny: Men's Perceptions of Female Power in Dating Relationships." PhD diss., University of Sussex.

artlone. 2014. Comment #4 on "Gone Girl." *A Voice for Men,* Entertainment Forum, October 5. https://d2ec906f9aea-003845.vbulletin.net/forum

/avfm-central/entertainment/13048-gone-girl?13232-Gone-Girl
=&langid=2.

Ashley, W. 2014. "The Angry Black Woman: The Impact of Pejorative Stereotypes on Psychotherapy with Black Women." *Social Work in Public
Health* 29: 27–34.

AsianAway. 2015. "Comprehensive Red Pill Books—Redux." *The Red Pill.*
Reddit. www.reddit.com/r/TheRedPill/comments/2mcokc/comprehensive
_red_pill_books_redux.

Asmis, E. 1982. "Lucretius' Venus and the Stoic Zeus." *Hermes* 110: 458–470.

———. 1996. "The Stoics on Women." In J. Ward 1996, 68–92.

Bacarisse, B. 2017. "The Republican Lawmaker Who Secretly Created Reddit's
Women-Hating 'Red Pill.'" *The Daily Beast*, April 24.

Baker, K. 2013. "Cockblocked by Redistribution: A Pick-up Artist in Denmark."
Dissent 60: 8–11.

Barchiesi, A. 1997. *The Poet and the Prince: Ovid and Augustan Discourse.*
Berkeley: University of California Press.

———. 2007. "Women on Top: Livia and Andromache." In Gibson, Green, and
Sharrock 2007, 96–120.

Barnes, J. 1997. *Logic and the Imperial Stoa.* Leiden: Brill.

Beale, T. (Vox Day) 2016. "(((Cathy Young))) Critiques the #AltRight." *Vox
Popoli*, August 28. https://voxday.blogspot.ca/2016/08/cathy-young
-critiques-altright.html.

Beard, M. 2014. "How Stoical Was Seneca?" *The New York Review of Books*,
October 2.

———. 2017. "The Latin Right." *The Times Literary Supplement*, March 31.

Bilsker, D. and J. White. 2011. "The Silent Epidemic of Male Suicide." *British
Columbia Medical Journal* 53: 529–534.

Black Label Logic. 2016. "Seneca and Machiavelli: Brothers in Arms." *Black
Label Logic*, May 20. https://blacklabellogic.com/2016/05/20/seneca-and
-machiavelli-brothers-in-arms/.

Blake, M. 2015. "Mad Men: Inside the Men's Rights Movement—and the Army
of Misogynists and Trolls It Spawned." *Mother Jones*, January/February.

Bloom, A. 1987. *The Closing of the American Mind: How Higher Education Has
Failed Democracy and Impoverished the Souls of Today's Students.* New
York: Simon and Schuster.

References

Bobzien, S. 1996. "Stoic Syllogistic." *Oxford Studies in Ancient Philosophy* 14: 133–92.

———. 1997. "The Stoics on Hypotheses and Hypothetical Arguments." *Phronesis* 42: 299–312.

———. 1999. "Logic: The Stoics." In K. Algra et al. *The Cambridge History of Hellenistic Philosophy*, 92–157. Cambridge: Cambridge University Press.

Bokhari, A. and M. Yiannopoulos. 2016. "An Establishment Conservative's Guide to the Alt-Right." *Breitbart News*, March 29. www.breitbart.com /tech/2016/03/29/an-establishment-conservatives-guides-to-the-alt-right/.

Bond, S. 2017. "Why We Need to Start Seeing the Classical World in Color." *Hyperallergic*, June 7.

Boyle, A. 1997. *Tragic Seneca: An Essay on the Rhetorical Tradition*. New York: Routledge.

Braund, S. 1992. "Juvenal—Misogynist or Misogamist?" *Journal of Roman Studies* 82: 71–86.

Braund, S. and G. Most, eds. 2007. *Ancient Anger: Perspectives from Homer to Galen*. Cambridge: Cambridge University Press.

Brouwer, R. 2014. *The Stoic Sage: The Early Stoics on Sagehood, Wisdom, and Socrates*. Cambridge: Cambridge University Press.

Brownmiller, S. 1975. *Against Our Will: Men, Women, and Rape*. New York: Fawcett.

Burns, P. 2014. "The Original PUA: Learn to Pick Up Chicks the Way the Ancient Romans Did." *Renaissance Man Journal*, November 5. https:// gainweightjournal.com/the-original-pua-learn-to-pick-up-chicks-the -way-the-ancient-romans-did/.

———. 2015. "Lessons from PUA Ovid: The Original Latin Lover." *Return of Kings*, January 23. www.returnofkings.com/53677/lessons-from-pua-ovid -the-original-latin-lover.

Burt, M. 1980. "Cultural Myths and Supports of Rape." *Journal of Personality and Social Psychology* 38: 217–30.

Bustillos, M. 2013. "Ken Hoinsky on Seduction, Women and Mistakes." *The Awl*, June 24.

Cahoon, L. 1988. "The Bed as Battlefield: Erotic Conquest and Military Metaphor in Ovid's *Amores*." *Transactions of the American Philological Association* 118: 293–307.

References

Cairns, D. 1993. *Aidōs: The Psychology and Ethics of Honour and Shame in Ancient Greek Literature*. Oxford: Oxford University Press.

Campbell, A. and K. Baker. 2016. "Unfounded: When Detectives Dismiss Rape Reports before Investigating Them." *Buzzfeed*, September 8. www.buzzfeed .com/alexcampbell/unfounded?utm_term=.mrWbLoBgN#.ioBY6w5LA.

Case, S. 1985. "Classic Drag: The Greek Creation of Female Parts." *Theater Journal* 37: 317–327.

Ceporina, M. 2012. "The *Meditations*." In *A Companion to Marcus Aurelius*, edited by van Ackeren, 45–61. Malden, MA and Oxford: Wiley-Blackwell.

Chubbs, B. 2014. "There Is Little Difference between Women throughout History." *Return of Kings*, April 29. www.returnofkings.com/34258/there -is-little-difference-between-women-throughout-history.

Cleary, S. 2016. "Stoicism Now: Conversation with Massimo Pigliucci." *Blog of the American Philosophical Association*, December 21. https://blog .apaonline.org/2016/12/21/stoicism-now-conversation-with-massimo -pigliucci/

Coates, T. 2015. "The Black Family in the Age of Mass Incarceration." *The Atlantic*, October.

Cohen, E. 2014. "Sexual Abuse and Sexual Rights: Slaves' Erotic Experience at Athens and Rome." In Hubbard 2014, 184–198.

Connolly, B. 2016. "Our Sovereign Father, Donald Trump." *Los Angeles Review of Books*, April 5.

Cooper, J. 1989. "Greek Philosophers on Euthanasia and Suicide." In *Suicide and Euthanasia (Philosophy and Medicine 35)*, edited by B. Brody, 9–38. Dordrecht, Boston: Kluwer Academic Publishers.

Craik, E. 1993. "ΑΙΔΩΣ in Euripides' *Hippolytos* 373–430: Review and Reinterpretation." *Journal of Hellenic Studies* 113: 45–59.

Crenshaw, K. 1992. "Whose Story Is It Anyway? Feminist and Antiracist Appropriations of Anita Hill." In *Race-ing Justice, En-gendering Power*, edited by T. Morrison, 402–440. New York: Pantheon.

Crowley, M. 2017. "Why the White House Is Reading Greek History." *Politico*, June 21.

Curtius, Q. 2015. *Thirty Seven: Essays on Life, Wisdom, and Masculinity*. Self-published through Amazon's CreateSpace.

———. 2016a. "The Details on My Upcoming Book *On Duties.*" *Quintus Curtius: Fortress of the Mind*, June 17. https://qcurtius.com/2016/06/17/the-details -on-my-upcoming-book-on-duties/comment-page-1/.

———. 2016b. "When Education Does Not Mean Knowledge: The Case of Mark Zuckerberg's Sister." *Quintus Curtius: Fortress of the Mind*, December 18. https://qcurtius.com/2016/12/18/a-response-to-a-detractor/.

Daniels, J. 2009. *Cyber Racism: White Supremacy Online and the New Attack on Civil Rights*. Lanham, MD: Rowman and Littlefield.

Darby, S. 2017. "The Rise of the Valkyries." *Harper's Magazine*, September.

Davidson, J. 2007. *The Greeks and Greek Love: A Radical Reappraisal of Homosexuality in Ancient Greece*. London: Weidenfeld and Nicolson.

De Beauvoir, S. 1948. *The Ethics of Ambiguity*. New York: Citadel.

De Ste. Croix, G. E. M. 1981. *The Class Struggle in the Ancient Greek World*. Ithaca: Cornell University Press.

Deacy, S. and K. Pierce, eds. 1997. *Rape in Antiquity: Sexual Violence in the Greek and Roman Worlds*. London: Duckworth.

Dee, J. 2003–4. "Black Odysseus, White Caesar: When Did 'White People' Become 'White'?" *CJ* 99: 157–167.

Denes, A. 2011. "Biology as Consent: Problematizing the Scientific Approach to Seducing Women's Bodies." *Women's Studies International Forum* 34: 411–9.

Denning, D., Y. Conwell, D. King, and C. Cox. 2000. "Method Choice, Intent, and Gender in Completed Suicide." *Suicide and Life-Threatening Behavior* 30: 282–288.

Dewey, C. 2014. "The Only Guide to Gamergate You Will Ever Need to Read." *Washington Post*, October 14.

Dickison, S. 1973. "Abortion in Antiquity." *Arethusa* 6: 159–166.

DiKaiomata, A. 2015. "Yes All Feminists Are Like That," *A Voice for Men*, February 18. www.avoiceformen.com/feminism/yes-all-feminists-are-like-that/.

Dover, K. 1989. *Greek Homosexuality*. Cambridge: Harvard University Press.

Dreher, R. 2016. "Re-Tribalizing America." *The American Conservative*, May 25.

DuBois, P. 2001. *Trojan Horses: Saving Classics from the Conservatives*. New York: New York University Press.

Dworkin, A. 2006. *Intercourse*. New York: Basic Books.

Eaton80. 2015. "Hell Has Frozen Over: Jezebel on False Rape Accusations." *MensRights*, Reddit. www.reddit.com/r/MensRights/comments/3f8lcr /hell_has_frozen_over_jezebel_on_false_rape/.

Elam, P. 2010. "Jury Duty at a Rape Trial? Acquit!" *A Voice for Men*, July 20. www.avoiceformen.com/mens-rights/jury-duty-at-a-rape-trial-acquit/.

———. 2012. "Adios, C-ya, Good-bye Man-o-sphere." *A Voice for Men*, September 5. https://www.avoiceformen.com/men/adios-man-o-sphere/.

Ellison, J. 2016. Letter to "Class of 2020 Student," University of Chicago, August. Published in facsimile in "U Chicago to Freshmen: Don't Expect Safe Spaces," by S. Jaschik. *Inside Higher Ed*, August 25, 2016. www .insidehighered.com/news/2016/08/25/u-chicago-warns-incoming -students-not-expect-safe-spaces-or-trigger-warnings.

Engel, D. 2003. "Women's Role in the Home and the State: Stoic Theory Reconsidered." *Harvard Studies in Classical Philology* 101: 267–288.

Epstein, J. 2010. "Male Studies vs. Men's Studies." *Inside Higher Ed*, April 8.

Erdely, S. 2014. "A Rape on Campus." *Rolling Stone* 1223 (December 4).

Esmay, D. 2016. "*To Kill a Mockingbird*: All Men Are Tom Robinson Now." *A Voice for Men*, February 19. www.avoiceformen.com/mens-rights/to -kill-a-mockingbird-all-men-are-tom-robinson-now/.

Fahrenthold, D. 2016. "Trump Recorded Having Extremely Lewd Conversation about Women in 2005." *The Washington Post*, October 7. www .washingtonpost.com/politics/trump-recorded-having-extremely-lewd -conversation-about-women-in-2005/2016/10/07/3b9ce776-8cb4-11e6-bf8a -3d26847eeed4_story.html?utm_term=.41c2876437ed.

Farrell, W. 1993. *The Myth of Male Power: Why Men Are the Disposable Sex*. New York: Simon and Schuster.

Fear, T. 2000. "The Poet as Pimp: Elegiac Seduction in the Time of Augustus." *Arethusa* 33: 151–158.

Ferrari, G. 2002. *Figures of Speech: Men and Maidens in Ancient Greece*. Chicago: University of Chicago Press.

Fisher, N. 1992. *Hybris: A Study in the Values of Honour and Shame in Ancient Greece*. London: Aris and Phillips.

Fisher, R. [pseud. redpillschool]. 2015. Comment on "Feminists Are Bitter Because They're Not Beautiful Enough to Be Raped like Other Women,"

References

by confessionberry. *The Blue Pill.* Reddit. www.reddit.com/r/TheBluePill
/comments/23848d/feminists_are_bitter_because_theyre_not_beautiful/.

Fitzgerald, G. 1973. "Misconception, Hypocrisy, and the Structure of Euripides'
Hippolytus." *Ramus* 2: 20–40.

Fleishman, G. 2013. "Kickstop: How a Sleazebag Slipped through Kickstarter's
Cracks." *BoingBoing,* June 22.

Forney, M. 2013. "The Case against Female Education." *Matt Forney,* December 2. https://mattforney.com/case-female-education/.

Foucault, M. 1978–86. *The History of Sexuality.* 3 vols. New York: Random House.

Frantzen, A. 2016. Excerpt from *The Boxer's Kiss: Men, Masculinity, and Femfog.*
Personal website (www.allenjfrantzen.com), January. Available at
http://archive.is/2w3va.

Furioso, R. 2016. "What to Do if Police Are Questioning You about a Sexual
Encounter." *Return of Kings,* August 27. www.returnofkings.com/93929
/what-to-do-if-police-are-questioning-you-about-a-sexual-encounter.

Furley, W. 1996. "Phaidra's Pleasurable *Aidos* (Eur. *Hipp.* 380–7)." *Classical
Quarterly* 46: 84–90.

Futrelle, D. 2017. "Inside the Dangerous Convergence of Men's-Rights Activists
and the Extreme Alt-Right." *New York Magazine,* March 31.

Gaca, K. 2014. "Martial Rape, Pulsating Fear, and the Sexual Maltreatment of
Girls (παῖδες), Virgins (παρθένοι), and Women (γυναῖκες) in Antiquity."
American Journal of Philology 135: 303–357.

Gaisser, J. 2002. "The Reception of Classical Texts in the Renaissance." In *The
Italian Renaissance in the Twentieth Century,* edited by A. J. Grieco, M.
Rocke, and F. Gioffredi Superbi, 387–400. Florence: Leo S. Olschki.

Galbi, D. 2010a. "More on Ovid and Roman Love Elegy." *Purple Motes,*
March 28. www.purplemotes.net/2010/03/28/more-on-ovid-and-roman
-love-elegy/.

———. 2010b. "Understanding Ovid's Satirical Roman Love Elegy." *Purple
Motes,* February 14. www.purplemotes.net/2010/02/14/understanding
-ovids-satirical-roman-love-elegy/.

Galinsky, K. 1992. *Classical and Modern Interactions: Postmodern Architecture,
Multiculturalism, Decline, and Other Issues.* Austin: University of Texas.

Gamel, M. 1989. "*Non Sine Caede:* Abortion Politics and Poetics in Ovid's
Amores." *Helios* 16: 183–206.

References

Gardner, H. 2012. "Ventriloquizing Rape in Menander's *Epitrepontes*." *Helios* 39: 121–143.

Gibert, J. 1997. "Euripides' *Hippolytus* Plays: Which Came First?" *Classical Quarterly* 47: 85–97.

Gibson, R. 1998. "Meretrix or Matrona? Stereotypes in Ovid, *Ars Amatoria* 3." *Papers of the Leeds Latin Seminar* 10: 295–312.

———. 1999. "Ovid on Reading: Reading Ovid. Reception in Ovid *Tristia* II." *JRS* 89: 19–37.

———, S. Green, and A. Sharrock, eds. 2007. *The Art of Love: Bimillennial Essays on Ovid's* Ars Amatoria *and* Remedia Amoris. Oxford: Oxford University Press.

Gill, C. 1988. "Personhood and Personality: The Four-Personae Theory in Cicero, *De Officiis* I." *Oxford Studies in Ancient Philosophy* 6: 169–199.

———. 2007. "Marcus Aurelius' *Meditations:* How Stoic and How Platonic?" In *Platonic Stoicism—Stoic Platonism: The Dialogue between Platonism and Stoicism in Antiquity*, edited by M. Bonazzi and C. Helmig, 189–207. Leuven, Belgium: Leuven University Press.

Gloyn, E. 2013. "Reading Rape in Ovid's *Metamorphoses:* A Test-Case Lesson." *Classical World* 106: 676–681.

Gold, B. ed. 2012. *A Companion to Roman Love Elegy*. Malden, MA and Oxford: Wiley-Blackwell.

Goldhill, O. 2016. "Silicon Valley Tech Workers Are Using an Ancient Philosophy Designed for Greek Slaves as a Life Hack." *Quartz*, December 17.

Grant, D. 2016. "The Classics and White Supremacy: A Response to Donna Zuckerberg." *Social Matter*, December 20. www.socialmatter.net/2016/12/20/classics-white-supremacy-response-donna-zuckerberg/.

Graver, M. 1998. "The Manhandling of Maecenas: Senecan Abstractions of Masculinity." *American Journal of Philology* 119: 607–632.

Greene, E. 2000. "Gender Identity and the Elegiac Hero in Propertius 2.1." *Arethusa* 33: 241–261.

———. 2012. "Gender and Elegy." In Gold 2012, 357–371.

Greene, R. 1998. *The 48 Laws of Power*. New York: Penguin.

Hadot, P. 1995. *Philosophy as a Way of Life*. Malden, MA and Oxford: Wiley-Blackwell.

———. 1998. *The Inner Citadel: The Meditations of Marcus Aurelius*. Trans. M. Chase. Cambridge: Harvard University Press.

Hall, E. 2008. "Putting the Class into Classical Reception." In *A Companion to Classical Receptions*, edited by L. Hardwick and C. Stray, 386–398. Malden, MA and Oxford: Wiley-Blackwell.

———. 2015. "Why I Hate the Myth of Phaedra and Hippolytus." *The Edithorial*, May 24. http://edithorial.blogspot.com/2015/05/why-i-hate-myth-of -phaedra-and.html.

Hall, J. 2002. *Hellenicity: Between Ethnicity and Culture.* Chicago: University of Chicago Press.

Hallett, J. and M. Skinner, eds. 1997. *Roman Sexualities.* Princeton: Princeton University Press.

Halperin, D. 1990. *One Hundred Years of Homosexuality.* New York: Routledge.

———, J. Winkler, and F. Zeitlin, eds. 1990. *Before Sexuality: The Construction of Erotic Experience in the Ancient Greek World.* Princeton: Princeton University Press.

Hanink, J. 2017. *The Classical Debt: Greek Antiquity in an Era of Austerity.* Cambridge: Harvard University Press.

Hanson, V. and J. Heath 1998. *Who Killed Homer?: The Demise of Classical Education and the Recovery of Greek Wisdom.* New York: Free Press.

Harris, E. 1990. "Did the Athenians Regard Seduction as a Worse Crime than Rape?" *Classical Quarterly* 40: 370–377.

———. 1997. Review of *Rape in Antiquity: Sexual Violence in the Greek and Roman Worlds,* by S. Deacy and K. Pierce, eds. *Échos du Monde Classique/Classical Views* 40: 483–496.

———. 2004. "Did Rape Exist in Classical Athens? Further Reflections on the Laws about Sexual Violence." *Dike* 7: 41–83.

Harrison, T. 1997. "Herodotus and the Ancient Greek Idea of Rape." In Deacy and Pierce 1997, 185–208.

Hart, A. 2016. "Voxplaining the Alt Right." *American Renaissance*, April 21.

Hayward, J. 2015. "Campus Special Snowflakes Melt upon Contact with Greek Mythology." *Breitbart*, May 12. www.breitbart.com/big-government/2015 /05/12/campus-special-snowflakes-melt-upon-contact-with-greek -mythology/.

Hejduk, J. 2014. Introduction to *The Offense of Love:* Ars Amatoria, Remedia Amoris, *and* Tristia 2, in Ovid 2014, 3–48.

Herman, J. 1992. *Trauma and Recovery: The Aftermath of Violence—From Domestic Abuse to Political Terror.* New York: Basic Books.

Hexter, R. 2007. "Sex Education: Ovidian Erotodidactic in the Classroom." In Gibson, Green, and Sharrock 2007, 298–317.

Hill, L. 2001. "The First Wave of Feminism: Were the Stoics Feminists?" *History of Political Thought* 22: 13–40.

"The History of Pickup and Seduction, Part I." 2016. *Pickup Culture*, March 17. https://pickupculture.com/2016/03/17/the-history-of-pickup-and -seduction-pt-1-the-pre-60s/.

Hobbes, T. 2016. "International Outrage: Men around the World Try to Stop Make-Up Artist Lecture Tour." *Return of Kings*, March 15. www.returnof kings.com/81845/international-outrage-men-around-the-world-try-to -stop-make-up-artist-lecture-tour.

Hoinsky, K. [pseudo. TofuTofu]. 2013. Reddit post, "Above the Game: Intro and My Story (Preview of My Upcoming Seduction Guide)." *Seddit.* Reddit. www.reddit.com/r/seduction/comments/11ng7n/above_the_game_intro _my_story_preview_of_my/.

Holiday, R. 2012. *Trust Me, I'm Lying: Confessions of a Media Manipulator.* New York: Penguin.

——. 2014. *The Obstacle Is the Way: The Timeless Art of Turning Trials into Triumph.* New York: Penguin.

——. 2016a. *Ego Is the Enemy.* New York: Penguin.

——. 2016b. "How Dr. Drew Pinsky Changed My Life." *Ryan Holiday: Meditations on Strategies and Life*, February 1. https://ryanholiday.net /how-dr-drew-pinsky-changed-my-life/.

hooks, b. 1995. *Killing Rage: Ending Racism.* New York: Henry Holt.

Hubbard, T., ed. 2014. *A Companion to Greek and Roman Sexualities.* Malden, MA and Oxford: Wiley-Blackwell.

Hughes, R. 2011. *Rome: A Cultural, Visual, and Personal History.* New York: Knopf.

Hutchinson, G. O. 2004. "Euripides' Other *Hippolytus.*" *Zeitschrift für Papyrologie und Epigraphik* 149: 15–28.

Inwood, B., ed. 2003. *The Cambridge Companion to the Stoics.* Cambridge: Cambridge University Press.

Irvine, W. 2008. *A Guide to the Good Life: The Ancient Art of Stoic Joy*. Oxford: Oxford University Press.

"Is MGTOW the Idea of Ancient Stoicism Repeating Itself?" 2015. *Rex Patriarch*, January 14. http://rexpatriarch.blogspot.com/2015/01/is-mgtow-idea-of -ancient-stoicism.html.

James, S. 2003. *Learned Girls and Male Persuasion: Gender and Reading in Roman Love Elegy*. Berkeley: University of California Press.

———. 2012. "Teaching Rape in Roman Elegy, Part II." In Gold 2012, 549–557.

———. 2014a. "Reconsidering Rape in Menander's Comedy and Athenian Life: Modern Comparative Evidence." In *Menander in Contexts*, edited by A. Sommerstein, 24–39. New York: Routledge.

———. 2014b. "Talking Rape in the Classics Classroom: Further Thoughts." In Rabinowitz and McHardy 2014, 171–186.

Jansen, C. 2015. "Viewing Stoicism from the Right." *Radix Journal*, June 23. https://www.radixjournal.com/2015/06/2015-6-23-viewing-stoicism-from -the-right/.

Jeffries, R. 1992. *How to Get the Women You Desire into Bed*. Self-published.

Jha, R. 2015. "There's Now a Campaign to End Discrimination against Men, or 'Mancrimination.'" *Buzzfeed India*, June 16.

Johnson, K., T. Lynch, E. Monroe, and T. Wang. 2015. "Our Identities Matter in Core Classrooms." *Columbia Spectator*, April 30.

Johnson, M. 2016. Introduction, notes, and translation to Ovid 2016.

Johnson, W. R. 1996. "Male Victimology in Juvenal 6." *Ramus* 25: 170–186.

Just, R. 1989. *Women in Athenian Law and Life*. New York: Routledge.

Kahn, M. 2005. *Why Are We Reading Ovid's Handbook on Rape? Teaching and Learning at a Women's College*. Boulder: Paradigm Publishers.

Kanin, E. 1994. "False Rape Allegations." *Archives of Sexual Behavior* 23: 81–92.

Kaster, R. 2007. Review of *Roman Manliness: Virtus and the Roman Republic*, by M. McDonnell. *BMCR*, February 28.

Kavi, W. 2015. "Feminism Comes Full Circle into Embracing Aristotle's 'Natural Slavery.'" *Return of Kings*, February 3. www.returnofkings.com/54181 /feminism-comes-full-circle-into-embracing-aristotles-natural-slavery.

Kendall, L. 2002. *Hanging Out in the Virtual Pub: Masculinities and Relation-ships Online*. Berkeley and Los Angeles: University of California Press.

References

Kendall, S. 2014. "An Open Letter to Asian Men of the West." *Return of Kings*, June 4. www.returnofkings.com/36359/an-open-letter-to-asian-men-of-the-west.

Ker, J. 2013. *The Deaths of Seneca.* Oxford: Oxford University Press.

Khan, I. 2010. "The Misandry Bubble." *The Futurist*, January 1. www .singularity2050.com/2010/01/the-misandry-bubble.html.

Kimmel, M. 2013. *Angry White Men: American Masculinity at the End of an Era.* New York: Nation Books.

Knox, B. 1992. "The Oldest Dead White European Males." *The New Republic*, May 25. https://newrepublic.com/article/77364/the-oldest-dead-white -european-males.

———. 1993. *The Oldest Dead White European Males: And Other Reflections on the Classics.* New York: Norton.

Kohn, T. 2008. "The Wishes of Theseus." *Transactions of the American Philological Association* 138: 379–392.

Kolbert, E. 2015. "Such a Stoic." *The New Yorker*, February 2.

Kovacs, D. 1980. "Shame, Pleasure, and Honor in Phaedra's Great Speech (Euripides, *Hippolytus* 375–87)." *American Journal of Philology* 101: 287–303.

Krauser, N. n.d. Promotional webpage for *Daygame. Krauser PUA.* https:// krauserpua.com/daygame-nitro-street-pick-up-for-alpha-males/.

———. 2014. *Daygame Mastery.* Self-published through Lulu.

Kupers, T. 2005. "Toxic Masculinity as a Barrier to Mental Health Treatment in Prison." *Journal of Clinical Psychology* 61: 713–724.

Lear, A. 2014. "Ancient Pederasty: An Introduction." In Hubbard 2014, 102–127.

Leonid, C. 2017. "The Problem of False Rape Accusations Is Not Going Away." *Return of Kings*, April 1. www.returnofkings.com/116992/the-problem-of -false-rape-accusations-is-not-going-away.

Lisak, D. 2011. "False Allegations of Rape: A Critique of Kanin." *Sexual Assault Report* 11: 1–2, 6, 9.

Lisak, D., L. Gardinier, S. Nicksa, and A. Cote. 2010. "False Allegations of Sexual Assault: An Analysis of Ten Years of Reported Cases." *Violence Against Women* 12: 1318–1334.

Liveley, G. 2012. "Teaching Rape in Roman Elegy, Part I." In Gold 2012, 541–548.

Long, A. A. and D. N. Sedley, eds. 1987. *The Hellenistic Philosophers.* Cambridge: Cambridge University Press.

Lorde, A. 1981. "The Uses of Anger: Women Responding to Racism." Keynote Address at the National Women's Studies Association Conference, June. Storrs, Connecticut. Published in *Sister Outsider: Essays & Speeches by Audre Lorde* (Berkeley: Crossing Press, 2007), 124–133.

Losemann, V. 1977. *Nationalsozialismus und Antike: Studien zur Entwicklung des Faches Alte Geschichte 1933–1945.* Hamburg: Historische Perspektive.

———. 2007. "Classics in the Second World War." In *Nazi Germany and the Humanities: How German Academics Embraced Nazism,* edited by W. Bialas and A. Rabinbach, 306–340. London: Oneworld.

Lubchansky, M. 2014. *Please Listen to Me.* www.listen-tome.com/save-me/.

Lukianoff, G. and J. Haidt. 2015. "The Coddling of the American Mind." *The Atlantic,* September 2015.

MacKinnon, C. 1987. *Feminism Unmodified: Discourses on Life and Law.* Cambridge: Harvard University Press.

———. 1989. *Toward a Feminist Theory of the State.* Cambridge: Harvard University Press.

"Mailbag: June 2015." 2015. *Illimitable Men,* July 6. https://illimitablemen.com/2015/07/06/monthly-mailbag-june-2015/.

Manning, C. E. 1973. "Seneca and the Stoics on the Equality of the Sexes." *Mnemosyne* 26: 170–177.

Marchand, S. 1996. *Down from Olympus: Archaeology and Philhellenism in Germany, 1750–1970.* Princeton: Princeton University Press.

Max, T. 2006. *I Hope They Serve Beer in Hell.* New York: Citadel Press.

———. 2013. "About *The Mating Grounds.*" *The Mating Grounds.* http://thematinggrounds.com/about-mating-grounds/.

———. 2015. *Mate: Become the Man Women Want.* New York: Little, Brown and Company.

McArdle, M. 2015. "What We Don't Know about False Claims of Rape." *Bloomberg View,* June 4.

McAuley, M. 2012. "Specters of Medea: The Rhetoric of Stepmotherhood and Motherhood in Seneca's *Phaedra.*" *Helios* 39: 37–72.

McCoskey, D. 2012. *Race: Antiquity and Its Legacy.* London: I. B. Tauris.

McDermott, E. 2000. "Euripides' Second Thoughts." *Transactions of the American Philological Association* 130: 239–259.

McDonnell, M. 2006. *Roman Manliness: Virtus and the Roman Republic.* Cambridge: Cambridge University Press.

McGrath, C. 2011. "The Study of Man (or Males)." *The New York Times,* January 7.

Mette-Dittmann, A. 1991. *Die Ehegesetze des Augustus: Eine Untersuchung im Rahmen der Gesellschaftspolitik des Princeps.* Stuttgart: Franz Steiner Verlag.

Moses, D. 1993. "Livy's Lucretia and the Validity of Coerced Consent in Roman Law." In *Consent and Coercion to Sex and Marriage in Ancient and Medieval Societies,* edited by A. Laiou, 39–81. Washington, DC: Dumbarton Oaks Research Library and Collection.

Murnaghan, S. 1988. "How a Woman Can Be More Like a Man: The Dialogue between Ischomachus and His Wife in Xenophon's *Oeconomicus.*" *Helios* 15: 9–22.

Musonius. 2010. *Musonius Rufus: Lectures and Sayings.* Translated by C. King. Edited by W. Irvine. Self-published by William Irvine through Amazon's CreateSpace.

Myerowitz, M. 1985. *Ovid's Games of Love.* Detroit: Wayne State University Press.

Myerowitz-Levine, M. 2007. "Ovid's Evolution." In Gibson, Green, and Sharrock 2007, 252–275.

"The Myth of Female Rationality." 2016. *Illimitable Men,* February 1 and 9. https://illimitablemen.com/2016/02/01/the-myth-of-female-rationality -part-1/and https://illimitablemen.com/2016/02/09/the-myth-of-female -rationality-part-2/.

Nagle, A. 2017. *Kill All Normies: Online Culture Wars from 4chan and Tumblr to Trump and the Alt-Right.* Winchester, UK: Zero Books.

Nashrulla, T. 2015. "These Are the Indian Women Fighting for 'Men's Rights.'" *Buzzfeed,* November 30.

Naso, B. 2014. "Xenophon's 'The Economist' Holds Valuable Lessons on a Woman's Education." *Return of Kings,* November 6. www.returnofkings .com/44698/xenophons-the-economist-holds-valuable-lessons-on-a -womans-education.

Nietzsche, F. 1990. *Unmodern Observations / Unzeitgemasse Betrachtungen.* Translated and edited by W. Arrowsmith. New Haven and London: Yale University Press.

None-Of-You-Are-Real. 2016. "Triggered SJWs Have Successfully Gotten Ovid's 'Metamorphoses' Removed from the Syllabus for a Required Core Course at Colombia [*sic*] University." *GGFreeForAll*. Reddit.www.reddit.com/r /GGFreeForAll/comments/3tuglm/triggered_sjws_have_successfully _gotten_ovids/.

Noonan, P. 2015. "The Trigger-Happy Generation." *The Wall Street Journal*, May 22.

Nussbaum, M. 1994. *The Therapy of Desire*. Princeton: Princeton University Press.

——. 2002. "The Incomplete Feminism of Musonius Rufus." In *The Sleep of Reason*, edited by M. Nussbaum and J. Shivola, 283–325. Chicago: University of Chicago Press.

O'Connor, M. 2017. "The Philosophical Fascists of the Gay Alt-Right." *New York Magazine*, April 30.

Omitowoju, R. 2002. *Rape and the Politics of Consent in Classical Athens*. Cambridge: Cambridge University Press.

Ortiz, J. 2015. "Hear Them Roar: Meet the Honey Badgers, the Women behind the Men's Rights Movement." *Marie Claire*, October.

Ovid. 1997. *Tales from Ovid*. Translated by T. Hughes. New York: Farrar, Straus and Giroux.

——. 2014. *The Offense of Love: Ars Amatoria, Remedia Amoris, and* Tristia 2. Translated by J. Hejduk. Madison: University of Wisconsin Press.

——. 2016. *Ovid on Cosmetics:* Medicamina Faciei Femineae *and Related Texts*. Edited and translated by M. Johnson. Oxford: Oxford University Press.

Packman, Z. 1993. "Call It Rape: A Motif in Roman Comedy and Its Suppression in English-Speaking Publications." *Helios* 20: 42–55.

Padilla Peralta, D. 2015. "From Damocles to Socrates: The Classics in / of Hip-Hop." *Eidolon*, June 8.

Painter, N. 2010. *The History of White People*. New York: Norton.

Parker, H. 2007. "Free Women and Male Slaves, or Mandingo Meets the Roman Empire." In *Fear of Slaves—Fear of Enslavement in the Ancient Mediterranean / Peur de l'esclave—Peur de l'esclavage en Mediterranee ancienne (Discours, représentations, pratiques)*, edited by A. Serghidou, 281–298. Franche-Comté: Presses Universitaires de Franche-Comté.

Patai, D. and N. Koertge. 1994. *Professing Feminism: Cautionary Tales from the Strange World of Women's Studies*. New York: Basic Books.

Patterson, C. 1991. "Marriage and the Married Woman in Athenian Law."
   In *Women's History and Ancient History*, edited by S. Pomeroy, 48–72.
   Chapel Hill: University of North Carolina Press.

Pigliucci, M. 2015. "How to Be a Stoic." *New York Times*, February 2.

PlainEminem. 2015. "Women Today Are Just Like Women in Ancient Rome."
   The Red Pill. Reddit. www.reddit.com/r/TheRedPill/comments/2254lc
   /women_today_are_just_like_women_in_ancient_rome/.

Pomeroy, S. 1974. "Feminism in Book V of Plato's *Republic*." *Apeiron* 8: 33–35.

Porter, J. 2007. "Adultery by the Book: Lysias 1 (*On the Murder of Eratosthenes*)
   and Comic *Diegesis*." In *Oxford Readings in Classical Studies: The Attic
   Orators*, edited by E. Carawan, 60–88. Oxford: Oxford University Press.
   Revised from *Echos du Monde Classique/Classical Views* 16 (1997): 421–53.

Preston, J. 2010. "Prosthetic White Hyper-Masculinities and 'Disaster Education.'" *Ethnicities* 10: 331–343.

Prins, Y. 2017. *Ladies' Greek: Victorian Translations of Tragedy*. Princeton:
   Princeton University Press.

Rabinowitz, N. and F. McHardy, eds. 2014. *From Abortion to Pederasty:
   Addressing Difficult Topics in the Classics Classroom*. Columbus: Ohio
   State University Press.

"The Rationalization Hamster Is Now Immortal." 2011. *The Private Man*,
   December 12.

Redfield, J. 1985. "Herodotus the Tourist." *Classical Philology* 80: 97–118.

Rensin, E. 2015. "The Internet Is Full of Men Who Hate Feminism. Here's What
   They're Like in Person." *Vox*, February 5.

Richlin, A. 2014. *Arguments with Silence: Writing the History of Roman Women*.
   Ann Arbor: University of Michigan Press.

Riddle, J. 1992. *Contraception and Abortion from the Ancient World to the
   Renaissance*. Cambridge: Harvard University Press.

———. 1999. *Eve's Herbs: A History of Contraception and Abortion in the West*.
   Cambridge: Harvard University Press.

Robertson, D. 2010. *The Philosophy of Cognitive Behavioural Therapy: Stoic
   Philosophy as Rational and Cognitive Psychotherapy*. London: Karnac.

Roche, H. 2013. *Sparta's German Children: The Ideal of Ancient Sparta in the
   Royal Prussian Cadet-Corps, 1818–1920, and in National-Socialist Elite
   Schools (the Napolas), 1933–1945*. Swansea: The Classical Press of Wales.

Roisman, H. 1999a. *Nothing Is as It Seems: The Tragedy of the Implicit in Euripides'* Hippolytus. Boston: Rowman & Littlefield.

———. 1999b. "The *Veiled Hippolytus* and Phaedra." *Hermes* 127: 397–409.

———. 2005. "Women in Senecan Tragedy." *Scholia* ns 14: 72–88.

Romm, J. 2014. *Dying Every Day: Seneca at the Court of Nero.* New York: Random House.

Rosati, G. 2007. "The Art of *Remedia Amoris:* Unlearning to Love?" In Gibson, Green, and Sharrock 2007, 143–165.

Rosenmeyer, T. 1986. "Stoic Seneca." *Modern Drama* 29: 92–109.

Rosenstein, N. 2004. *Rome at War: Farms, Families, and Death in the Middle Republic.* Chapel Hill: University of North Carolina Press.

Russ, J. 1980. *On Strike against God: A Lesbian Love Story.* Brooklyn: Out & Out Books.

Rutz, W. and Z. Rihmer. 2007. "Suicidality in Men—Practical Issues, Challenges, Solutions." *Journal of Men's Health and Gender* 4: 393–401.

Saller, R. 1987. "Men's Age at Marriage and Its Consequences in the Roman Family." *Classical Philology* 82: 21–34.

Salway, B. 1994. "What's in a Name? A Survey of Roman Onomastic Practice from c. 700 BC to AD 700." *JRS* 84: 124–145.

Sambursky, S. 1987. *Physics of the Stoics.* Princeton: Princeton University Press.

Sandbach, F. 1989. *The Stoics.* London: Bristol Classical Press.

Sansone, D. 2016. "Herodotus on Lust." *Transactions of the American Philological Association* 146: 1–36.

Saunders, C. 2012. "The Truth, the Half-Truth, and Nothing Like the Truth: Reconceptualizing False Allegations of Rape." *British Journal of Criminology* 52: 1152–1171.

Savage, C. 2016. "6 Ways 'Misogynists' Do a Better Job at Helping Women than Feminists." *Return of Kings*, February 8. www.returnofkings.com/77012/6-ways-misogynists-do-a-better-job-at-helping-women-than-feminists.

Saxton, D. 2015. "An Invitation for SlutHaters to Join the Philosophy of Rape." SlutHate.com, June. http://sluthate.com/viewtopic.php?t=93095.

Schambelan, E. 2016. "Pseudo-Conservatism, the Soldier Male, and the Air Horn." *Los Angeles Review of Books*, April 18.

Scheidel, W. 2007. "Roman Funerary Commemoration and the Age at First Marriage." *CP* 102: 389–402.

Schmitz, R. and E. Kazak. 2016. "Masculinities in Cyberspace: An Analysis of Portrayals of Manhood in Men's Rights Activist Websites." *Social Sciences* 5, no. 2, 18.

Schofield, M. 1991. *The Stoic Idea of the City.* Cambridge: Cambridge University Press.

———. 2003. "Stoic Ethics." In Inwood 2003, 233–256.

Sealey, R. 1990. *Women and Law in Classical Greece.* Chapel Hill: University of North Carolina Press.

Sedley, D. 2003. "The School, From Zeno to Arius Didymus." In Inwood 2003, 7–32.

Segal, C. 1970. "Shame and Purity in Euripides' *Hippolytus*." *Hermes* 98: 278–299.

Serwer, A. and K. Baker. 2015. "How Men's Rights Leader Paul Elam Turned Being a Deadbeat Dad into a Moneymaking Movement." *Buzzfeed*, February 6.

Sharpe, D. 2014. "The Pros and Cons of Game for Black Men in the West." *Return of Kings*, November 25. www.returnofkings.com/47823/the-pros-and-cons-of-game-for-black-men-in-the-west.

———. 2016. "5 Tips for Non-White Men Who Want to Date White Women." *Return of Kings*, April 27. http://www.returnofkings.com/85719/5-tips-for-non-white-men-who-want-to-date-white-women.

Sharrock, A. 1994. *Seduction and Repetition in Ovid's* Ars Amatoria II. Oxford: Clarendon Press.

———. 2007. "Love in Parentheses: Digression and Narrative Hierarchy in Ovid's Erotodidactic Poems." In Gibson, Green, and Sharrock 2007, 23–39.

Sims, J. 2010. "What Race Were the Greeks and Romans?" *American Renaissance*, October.

Smith, A. [pseudo. Chrysoberyl]. 2012. "Train Game—Fun for Everyone." *Real Social Dynamics*, October 25. Available at http://web.archive.org/web/20130109110015/http:/www.rsdnation.com/node/249698/forum?

Smith, B. 2016. "The Straight Men Who Want Nothing to Do with Women." *MEL Magazine*, September 28.

Smith, J. 2014. "Gone Girl's Recycling of Rape Myths Is a Disgusting Distortion." *The Guardian*, October 6. www.theguardian.com/commentisfree/2014/oct/06/gone-girl-rape-domestic-violence-ben-affleck.

Sommerstein, A., ed. 1994. *Thesmophoriazusae,* by Aristophanes. Warminster: Aris and Phillips.

Strategos_autokrator. 2015. "How to Become Outcome Independent Using a Stoic Trick." *The Red Pill.* Reddit. www.reddit.com/r/TheRedPill/comments /2ruugd/how_to_become_outcome_independent_using_a_stoic/.

Strauss, N. 2005. *The Game: Penetrating the Secret Society of Pickup Artists.* New York: Harper Collins.

———. 2015. *The Truth: An Uncomfortable Book about Relationships.* New York: Harper Collins.

Sulprizio, C. 2015. "Why Is Stoicism Having a Cultural Moment?" *Eidolon,* August 17.

Supremo. 2016. "Why False Rape Is Far Worse than Rape." MGTOW, June 13. www.mgtow.com/forums/topic/why-false-rape-is-far-worse-than-rape/.

Swann, J. 2016. "Pick-Up Artists See a Kindred Spirit in Trump." *MEL Magazine,* November 30.

Taleb, N. N. 2012. *Antifragile: Things That Gain From Disorder.* New York: Random House.

Thakur, S. 2014. "Challenges in Teaching Sexual Violence and Rape: A Male Perspective." In Rabinowitz and McHardy 2014, 152–170.

Theweleit, K. 1977. *Männerphantasien.* Frankfurt: Verlag Roter Stern.

Thorn, C. 2012. *Confessions of a Pickup Artist Chaser: Long Interviews with Hideous Men.* Self-published through Amazon's CreateSpace.

Timberg, S. 2015. "How University Trigger Warnings Will Backfire: Does Fox News Need Any More Ammunition against the Humanities?" *Salon,* May 15.

Tomassi, R. 2013. "Anger Management." *The Rational Male,* November 6. https://therationalmale.com/2013/11/06/anger-management/.

———. 2014. "The Apologists." *The Rational Male,* April 28. https:// therationalmale.com/tag/stony-brook-university/.

———. 2017. "The Anger Bias." *The Rational Male,* March 29. https:// therationalmale.com/2017/03/29/the-anger-bias/.

Treggiari, S. 1982. "Consent to Roman Marriage: Some Aspects of Law and Reality." *CV/ECM* 26: 34–44.

———. 1991. *Roman Marriage: Iusti Coniuges from the Time of Cicero to the Time of Ulpian.* Oxford: Oxford University Press.

Tuthmosis. 2014. "What Really Happened to Tucker Max?" www.thumotic.com. No longer available online.

Valenti, J. 2014. "Choosing Comfort over Truth: What It Means to Defend Woody Allen." *The Nation*, February 3.

Valizadeh, R. n.d. Promotional page for *Bang* series.

———. 2008. "It Doesn't Matter if She Orgasms or Not." *Roosh V*, July 24. www .rooshv.com/it-doesnt-matter-if-she-orgasms-or-not.

———. 2010. "When No Means Yes." *Roosh V*, June 18. www.rooshv.com/when -no-means-yes.

———. 2011a. "5 Reasons Your Game Sucks." *Roosh V*, October 10. www.rooshv .com/5-reasons-your-game-sucks.

———. 2011b. *Bang Iceland*. Self-published through Amazon's CreateSpace.

———. 2011c. *Don't Bang Denmark*. Self-published through Amazon's CreateSpace.

———. 2011d. "More Book Reviews 9." *Roosh V*, December 28. www.rooshv.com /more-book-reviews-9.

———. 2011e. "You Did This to Me." *Roosh V*, October 19. www.rooshv.com/you -did-this-to-me.

———.2012a. *30 Bangs: The Shaping of One Man's Game from Patient Mouse to Rabid Wolf*. Self-published through Amazon's CreateSpace.

———. 2012b. *Bang Ukraine*. Self-published through Amazon's CreateSpace.

———. 2013. "10 Reasons Why Heterosexual Men Should Leave America." *Roosh V*, September 16. www.rooshv.com/10-reasons-why-heterosexual-men -should-leave-america

———. 2014a. "All Public Rape Allegations Are False." *YouTube*, December 7. www.youtube.com/watch?v=bySyocJzroE.

———. 2014b. "The Decimation of Western Women Is Complete." *Roosh V*, December 15. www.rooshv.com/the-decimation-of-western-women-is -complete.

———. 2014c. "What Is a Social Justice Warrior (SJW)?" *Roosh V*, October 6. www.rooshv.com/what-is-a-social-justice-warrior-sjw.

———. 2015a. "The Accusation That I'm a Rapist Is a Malicious Lie." *Roosh V*, November 11. www.rooshv.com/the-accusation-that-im-a-rapist-is-a -malicious-lie.

———. 2015b. "How to Stop Rape." *Roosh V*, February 15. www.rooshv.com/how -to-stop-rape.

———. 2015c. "An Introduction to Stoicism with the *Enchiridion* by Epictetus." *Return of Kings*, April 17. www.returnofkings.com/60570/an-introduction-to-stoicism-with-the-enchiridion-by-epictetus.

———. 2015d. "Men Should Start Recording Sex with a Hidden Camera." *Roosh V*, October 5. www.rooshv.com/men-should-start-recording-sex-with-a-hidden-camera.

———. 2015e. "The Principal Tenets of Stoicism by Seneca." *Return of Kings*, June 1. www.returnofkings.com/64452/the-principal-tenets-of-stoicism-by-seneca.

———. 2015f. "What Is Neomasculinity?" *Roosh V*, May 6. www.rooshv.com/what-is-neomasculinity.

———. 2015g. "Women Must Have Their Behavior and Decisions Controlled by Men." *Roosh V*, September 21. www.rooshv.com/women-must-have-their-behavior-and-decisions-controlled-by-men.

———. 2016a. "Marcus Aurelius' *Meditations* Is the Best Manual We Have on How to Live." *Roosh V*, March 2. www.rooshv.com/marcus-aurelius-meditations-is-the-best-manual-we-have-on-how-to-live.

———. 2016b. "What Donald Trump's Victory Means for Men." *Return of Kings*, November 11. www.returnofkings.com/100669/what-donald-trumps-victory-means-for-men.

———. 2017. "How to Save Western Civilization." *Roosh V*, March 6. www.rooshv.com/how-to-save-western-civilization.

Vlastos, G. 1997. "Was Plato a Feminist?" In *Plato's Republic: Critical Essays*, edited by R. Kraut, 115–128. New York: Rowman and Littlefield.

Vogt, K. 2006. "Anger, Present Justice and Future Revenge in Seneca's *De Ira*." In *Seeing Seneca Whole: Perspectives on Philosophy, Poetry and Politics*, edited by K. Volk and G. D. Williams, 57–74. Leiden: Brill.

Volk, K. 2007. "*Ars Amatoria Romana:* Ovid on Love as a Cultural Construct." In Gibson, Green, and Sharrock 2007, 235–251.

Von Markovik, E. 2007. *The Mystery Method: How to Get Beautiful Women into Bed*. New York: St. Martin's Press.

Wachowski, L. and L. Wachowski. 1999. *The Matrix*. Warner Brothers.

Walcot, P. 1978. "Herodotus on Rape." *Arethusa* 11: 137–147.

Walley-Jean, J. 2009. "Debunking the Myth of the 'Angry Black Woman': An Exploration of Anger in Young African American Women." *Black Women, Gender + Families* 3: 68–86.

Walters, J. 1997. "Invading the Roman Body: Manliness and Impenetrability in Roman Thought." In *Roman Sexualities*, edited by J. Hallett and M. Skinner, 29–46. Princeton: Princeton University Press.

Ward, C. 1995. *Attitudes toward Rape: Feminist and Social Psychological Perspectives*. London: SAGE.

Ward, I. 2013. "Rape and Rape Mythology in the Plays of Sarah Kane." *Comparative Drama* 47: 225–248.

Ward, J., ed. 1996. *Feminism and Ancient Philosophy*. New York: Routledge.

Watson, L. 2007. "The Bogus Teacher and His Relevance for Ovid's *Ars Amatoria*." *Rheinisches Museum* 150: 337–374.

Weaver, P. 1994. "Epaphroditus, Josephus, and Epictetus." *Classical Quarterly* 44: 468–479.

Weidmann, J. n.d. "Sixteen Commandments of Poon." *Chateau Heartiste*. https://heartiste.wordpress.com/the-sixteen-commandments-of-poon/.

———. 2013. "Recommended Great Books for Aspiring Womanizers." *Chateau Heartiste*, September 4. https://heartiste.wordpress.com/2013/09/04/recommended-great-books-for-aspiring-womanizers/.

———. 2016a. "A Hot White Woman's Gine Is a Terrible Thing to Waste." *Chateau Heartiste*, January 20. https://heartiste.wordpress.com/2016/01/20/a-hot-white-womans-gine-is-a-terrible-thing-to-waste/.

———. 2016b. "The Patented CH 'How normal are you?' Opener." *Chateau Heartiste*, May 23. https://heartiste.wordpress.com/2016/05/23/the-patented-ch-how-normal-are-you-opener/.

———. 2016c. "Shiv of the Week: Choices and Consequences." *Chateau Heartiste*, May 20. https://heartiste.wordpress.com/2016/05/20/shiv-of-the-week-choices-and-consequences/.

———. 2016d. "Tattoos as Maimgeld." *Chateau Heartiste*, May 6. https://heartiste.wordpress.com/2016/05/06/tattoos-as-maimgeld/.

———. 2017a. "Powerlust Moves." *Chateau Heartiste*, April 24. https://heartiste.wordpress.com/2017/04/24/powerlust-moves/.

———. 2017b. "Single White Women Want to Spread Their Legs for the World." *Chateau Heartiste*, May 11. https://heartiste.wordpress.com/2017/05/11/single-white-women-want-to-spread-their-legs-for-the-world/.

Weiner, R. 2016. "Titus in Space." *The Paris Review*, November 29.

Welch, T. 2005. "*Amor* versus Roma: Gender and Landcape in Propertius 4.4." In Ancona and Greene 2005, 296–317.

Wheatcroft, J. and S. Walklate. 2014. "Thinking Differently about 'False Allegations' in Cases of Rape: The Search for Truth." *International Journal of Criminology and Sociology* 3: 239–248.

White, M. 2003. "Stoic Natural Philosophy Physics and Cosmology." In Inwood 2003, 124–152.

Whitmarsh, T. 2015. *Battling the Gods: Atheism in the Ancient World*. New York: Knopf.

Williams, B. 1993. *Shame and Necessity*. Berkeley: University of California Press.

Wilson, E. 2014. *The Greatest Empire: A Life of Seneca*. Oxford: Oxford University Press.

Wolfe, T. 1998. *A Man in Full*. New York: Farrar, Straus and Giroux.

Wray, D. 2001. *Catullus and the Poetics of Roman Manhood*. Cambridge: Cambridge University Press.

——. 2012. "Catullus the Roman Love Elegist?" In Gold 2012, 25–38.

Wyke, M. 1987. "The Elegiac Woman at Rome." *PCPS* 213: 153–178.

——. 2007. *The Roman Mistress*. Oxford: Oxford University Press.

Xenophon. 1994. *Oeconomicus: A Social and Historical Commentary*. Edited and translated by S. Pomeroy. Oxford: Clarendon Press.

Yates, V. 2015. "Biology Is Destiny: The Deficiencies of Women in Aristotle's Biology and *Politics*." *Arethusa* 48: 1–16.

Young, C. 2014. "Woody Allen, Feminism, and 'Believing the Survivor.'" *Time*, February 12.

Zadrozny, B. 2016. "The Pickup Artist Rape Ring." *The Daily Beast*, September 21. www.thedailybeast.com/pickup-artists-preyed-on-drunk-women-brought-them-home-and-raped-them.

Zuckerberg, D. 2015. "He Said, She Said: The Mythical History of the False Rape Allegation" *Jezebel*, July 30. https://jezebel.com/he-said-she-said-the-mythical-history-of-the-false-ra-1720945752.

——. 2016. "How to Be a Good Classicist under a Bad Emperor." *Eidolon*, November 21.

——. 2017. "'Learn Some F*cking History.'" *Eidolon*, October 5.

Zvan, S. 2014. "But How Do You Know the MRAs Are Atheists?" *The Orbit*, April 13.

ACKNOWLEDGMENTS

I discovered the Red Pill community's interest in Classics in August 2015, when I looked into the traffic sources for an *Eidolon* article and came across an acrimonious debate on r / Stoicism about the Red Pill's obsession with that ancient philosophy. As that observation turned into an idea, then a book proposal, and then a book, I sought guidance from many teachers, colleagues, and friends. Almost every one of them told me that they were happier before they learned about this disturbing material, and that reading what I had written made them want to quit the internet. I am deeply grateful that they helped me anyway.

Helen Morales has been a font of indispensable wisdom and encouragement throughout the entire process: she gave notes on the first draft of the manuscript, dispensed frank advice, and, most important, introduced me to Sharmila Sen, my wonderful editor. Sharmila and her team at Harvard University Press, especially Heather Hughes, took a chance on me as a first-time author and believed in this book long before anybody else realized how timely it was. Their assistance and guidance have been instrumental. I am also grateful to the two anonymous reviewers, whose advice was an invaluable resource as I revised my manuscript after the 2016 election.

It takes a village to create a book, and many brilliant people offered suggestions on all or part of this manuscript. Sarah Scullin, David Kaufman, Tara Mulder, Tori Lee, and Ali Wunderman all read and gave advice on chapters of this book; so did Sharon James, who also kindly offered to let me virtually sit in on her Ovid seminar. Yung In Chae was my right hand throughout the writing of the first draft, simultaneously editing my manuscript and keeping *Eidolon* running. So many people shared their wisdom and helped shape my ideas that it would be impossible to name them all, but I especially want to recognize Randi Zuckerberg, who was unstintingly generous in sharing her experience with publishing, and Johanna Hanink and Dan-el Padilla Peralta, whose belief in the importance of this book constantly pushed me to do more.

Spending so much time reading and studying some of the darkest corners of the internet took a toll, and I'm deeply thankful to all those who kept my spirits up: my wonderful friends, especially Marina Danilevsky, Karyl Kopaskie, and Mallory Monaco; my therapist, who keeps me sane in a crazy world; Analisa Naldi and Sean Arnold, who taught me that there are no better mood-lifters than weightlifting and boxing; Harry, my beloved partner in all things; and Jonah, who fills my days with joy and our house with LEGO® bricks.

When I was in college, I took a class on the writings of J. R. R. Tolkien. One day, when talking about Tolkien's education, my professor said, "If you ever want to make your parents cry, tell them you want to become a philologist." I am endlessly grateful to my father, Edward Zuckerberg, for taking the news of my chosen career path without surprise or tears. Last, and most of all, I want to thank my mother, Karen Zuckerberg, who has undoubtedly read more words written by me than any other person on the planet and has been an unfailing source of love and support (and, more recently, much-needed childcare). She never stopped rooting for this project. This book is for her.

INDEX

Page numbers followed by n or nn indicate notes.

Golden Dawn political party, in
  Greece, 195n2
Gone Girl (film), 219n8
Grant, Sandy, 69
Greekness concept, 155, 211n30
Greene, Robert, 62
*Guide to the Good Life: The Ancient
  Art of Stoic Joy, A* (Irvine), 47
Gynocentrism, Red Pill commu-
  nity's beliefs about, 6, 11, 16–17,
  123, 146, 163–164, 172

Hall, Edith, 23, 179–181
Hamilton, Patrick, 39–40
"Hamstering," 31, 200n38
Hanson, Victor Davis, 36–37
Harpagē (abduction / theft), 155, 160
Harris, Edward, 163, 221n34, 223n53
"Having game," 17, 92–94, 100
Headphones, approaching women
  who are using, 109–110
Heath, John, 36–37
Hejduk, Julia, 89–90
Helen of Troy, abduction of,
  155–156, 157, 158, 221n31
Hellenistic Philosophers, The (Long
  and Sedley), 49
Herodotus of Halicarnassus,
  155–158
"Herodotus the Tourist" (Redfield),
  155
Heroides (Ovid), 34, 104, 129,
  168–169
Hesiod, 27–28, 29, 43, 72, 166, 187
Hill, Lisa, 70, 206n49

Hippolytus. *See* Phaedra and
  Hippolytus myths
Hippolytus (Euripides), 165–168,
  169, 170–171, 176, 181, 182, 223n55,
  225n61, 227n86
Hippolytus the Wreath-Bearer
  (Euripides), 224–225n60
Histories (Herodotus of Halicar-
  nassus), 155–158
Hoinsky, Ken "TofuTofu," 111–112
Holiday, Ryan, 61–65, 66, 68–69,
  79–80, 88, 205nn35,36
Homer, 22, 158, 221n28
Homosexuality, seduction tech-
  niques and, 124–128, 216n85,
  222n41
"Honey badgers," women as, 17
hooks, bell, 83
Horace, 128
"How to Be a Good Classicist under
  a Bad Emperor" (Zuckerberg),
  186, 188
"How to Save Western Civilization"
  (blog post), 177
Hybris (crime / outrageous act), 160
Hymn to Zeus (Cleanthes), 51
Hypergamy, 31, 94, 102–103, 200n38

Icarus, 98, 211n20
Identitarians, 20
Identity Evropa, 1, 20, 177
Ideological sublimation, as Red Pill
  tactic, 42–43
I Hope They Serve Beer in Hell
  (Max), 61, 101, 131

Life of Augustus (Suetonius), 215n715

"Live consistently" philosophy, 51, 81

Livy, 158–159, 161, 221n37

Logic (*logikē*), of Stoics, 49, 50, 52, 53, 203n10, 204n11

Logos: Stoics and, 63, 71, 73, 85–86, 203n5, 209–210n90. *See also* Rationality and emotional control

"Lolwut," 21, 198n20

Long, A., 49

Lorde, Audre, 83–84

Lubchansky, Matt, 199n30

Lucretia, rape of, 159, 162, 179, 222n39

Lyaéus, 113, 214n62

Lysias, 157, 161

MacKinnon, Catherine, 152, 153, 174

Male circumcision, 16

Male privilege, myth of, 12, 198n12

Man in Full, A (Wolfe), 202n3

Manosphere, use of term, 15

Marcus Aurelius, 58–59, 75; Holiday's philosophy and, 62–65; Red Pill community and, 9–10, 46, 60; Stoic philosophy and, 52–53, 65–66, 205n22

Marriage: classical tradition and, 28–31, 72, 199n34; consent requirements, 161–162, 223nn48, 49; MGTOW and strikes against, 19, 65; rape victims and, 160

Mate: Become the Man Women Want (Max), 131

Matrix, The (film), 1–2, 13

Max, Tucker, 61, 62, 131, 205n36

McArdle, M., 220n15

McAuley, M., 225n64, 226n68

McGill-Rutherford, Emily, 82

Meditations (Marcus Aurelius), 50, 58, 60–63, 74–75, 85

Men Going Their Own Way (MGTOW), 19–20, 29, 40, 44, 64–65, 123, 124, 144, 166

Meninism (masculism), 16, 31

"Men Should Start Recording Sex with a Hidden Camera" (blog post), 144

Men's human rights advocates (MHRAs), 16–17, 40

Men's human rights movement (MHRM), 17–18, 19, 42, 44, 198n12

Men's rights activists (MRAs), 15–16, 106, 144, 147

Meretrix, sexual relations and, 91–92, 117, 125, 128, 140, 210n7

Metamorphoses (Ovid), 33–35, 95, 141–142, 162, 168, 225n61

MGTOW.com, 19, 144

"MGTOW Manifesto," 29

"Misandry Bubble, The" (Khan), 41

Misogyny, 77, 185; classical traditions and, 6, 25, 27–29, 31–32, 70, 189; NAWALT and, 26; and sexism of Stoics, 68–76; social media and, 3, 14–15

Molōn labe (come and get them), 196n9